SpringerBriefs in Business

More information about this series at http://www.springer.com/series/8860

Marco Fortis · Monica Carminati

The Automatic Packaging Machinery Sector in Italy and Germany

Springer

Marco Fortis
Department of International Economics,
Institutions and Development
Università Cattolica del Sacro Cuore
Milan
Italy

Monica Carminati
Fondazione Edison
Milan
Italy

ISSN 2191-5482 ISSN 2191-5490 (electronic)
ISBN 978-3-319-12762-0 ISBN 978-3-319-12763-7 (eBook)
DOI 10.1007/978-3-319-12763-7

Library of Congress Control Number: 2014953921

Springer Cham Heidelberg New York Dordrecht London

Printed on acid-free paper

Springer is part of Springer Science+Business Media (www.springer.com)

Contents

Chapter 1
Introduction

Keywords Industrial districts · Packaging machinery industry · Emilia/Bologna packaging machinery district · Baden-Württemberg packaging machinery district

Industrial districts are one of the central pillars of the Italian economy. At the dawn of the new millennium the rise of the globalization and the advent of Asian competition brought profound changes to the production dynamics of the districts and their dominant model of specialization (Fortis 1998, 2000, 2001, 2004, 2005a, b, 2006, 2011; Fortis and Quadrio Curzio 2006; Fortis and Carminati 2007a, b, 2009, 2012; Quadrio Curzio and Fortis 2000, 2007, 2012).

Over the last decade in particular we have seen a contraction of the districts operating in the more traditional sectors, such as personal items and household goods (textiles-clothing, leather-footwear, furniture, tiles, etc.), and significant growth of medium and high-tech districts such as aircraft at Vergiate and Naples, pharmaceuticals in Latina and Frosinone, biomedical products in Mirandola, and electronics in the Etna Valley, to name but a few. Then there are the many districts in the mechanical engineering sector. Alongside the better known industrial segments such as industrial machinery, taps/valves, pumps and mechatronics, etc., a particularly interesting sector that combines traditional engineering with more advanced mechanics and electronics is that of the Emilia/Bologna automatic packaging machinery district. This case is especially interesting because it competes directly with a major industrial hub in an advanced economy, such as Germany, world leader in the machinery sector and the benchmark for competitiveness, where the state of Baden-Württemberg has one of the biggest concentrations of producers of automatic packaging machines.

The city of Bologna is especially well known worldwide for its high concentration of industrial companies specialized in the manufacture of packaging machinery. This is a strongly export-oriented sector, with a ratio of exports to total sales equal to 88.1 %, the highest among Italian manufacturers of capital goods. In October 2013 Unindustria Bologna launched "Globol", the first international forum dedicated to packaging, with Romano Prodi, the leader of the scientific committee.

The Emilia/Bologna district is a competitor of international importance, as Martin Schulz, German President of the European Parliament, made clear in an interview

M. Fortis and M. Carminati, *The Automatic Packaging Machinery Sector in Italy and Germany*, SpringerBriefs in Business,
DOI 10.1007/978-3-319-12763-7_1

with the magazine l'Espresso in May 2013: "*Italy has companies with great traditions and creative ability. Emilia is every bit as good as the Neckar valley, the heart of the Baden-Württemberg region*" (Vastano 2013).

This paper sets out to analyze the economy of the Emilia/Bologna wrapping and packaging machinery district and compare the most recent trends with those of its direct German rival, Baden-Württemberg. The Emilia/Bologna district, which spread out from the original province of Bologna to that of Modena, but can also be more broadly defined to include Parma and Reggio Emilia, employed over 16,000 workers in 2011 for a turnover of more than 3.7 billion euros. The German district, covering the adjacent towns of Schwäbisch Hall and Waiblingen in Baden-Württemberg to the north-east of Stuttgart, had an estimated 13,000 employees in 2010.

References

Fortis M (1998) Il made in Italy. Il Mulino, Bologna

Fortis M (2000) Il made in Italy nell'economia italiana e mondiale: il rilievo delle piccole e medie imprese e dei distretti. In: Quadrio Curzio A, Fortis M (eds) Il made in Italy oltre il 2000. Fondazione Edison Series, Il Mulino, Bologna

Fortis M (2001) Innovazione tecnologica e sistemi territoriali: specializzazioni produttive e interdipendenze. In Atti dei Convegni Lincei, no. 177, conference "Tecnologia e Società", volume I—Tecnologia, Produttività, Sviluppo, Accademia Nazionale dei Lincei, Rome

Fortis M (2004) La dinamica dei Sistemi locali e dei Distretti italiani: dal territorio ai mercati internazionali. In Garonna P and Gros-Pietro G M (eds) Il modello italiano di competitività, Ricerca per il Convegno Biennale 2004 del Centro Studi Confindustria, Edizioni Il Sole 24 Ore, Economia e Management

Fortis M (2005a) Il made in Italy nel "nuovo mondo". Protagonisti, Sfide, Azioni, Ministero delle Attività Produttive. Report prepared for the 2nd national conference on Foreign Trade, Rome, Feb 26

Fortis M (2005b) Le due sfide del made in Italy: globalizzazione e innovazione. Profili di analisi della Seconda Conferenza Nazionale sul commercio con l'estero. Fondazione Edison Series. Bologna, Il Mulino

Fortis M (2006) I distretti produttivi e la loro rilevanza nell'economia italiana: alcuni profili di analisi. In: Fortis M, Quadrio Curzio A (eds) Industria e distretti. Un paradigma di perdurante competitività italiana. Fondazione Edison Series, Bologna, Il Mulino

Fortis M (2011) Esportazioni, investimenti diretti esteri e competitività del sistema Paese nel mercato internazionale. In Economia italiana, 1/2011

Fortis M, Quadrio Curzio A (eds) (2006) Industria e distretti. Un paradigma di perdurante competitività italiana. Fondazione Edison Series, Bologna, Il Mulino

Fortis M, Carminati M (2007a) I distretti industriali nella concretezza economica e normativa: i «campioni territoriali» dell'Italia. In: Quadrio Curzio A, Fortis M (eds) Valorizzare un'economia forte. L'Italia e il ruolo della sussidiarietà. Fondazione Edison Series, Bologna, Il Mulino

Fortis M, Carminati M (2007b) Industrial districts: the economic reality and legislative framework in Italy. In Review of economic conditions in Italy, no. 1/2007

Fortis M, Carminati M (2009) Sectors of excellence in the Italian industrial districts. In: Becattini G, Bellandi M, De Propis L (eds) A hand-book of Industrial districts. Edward Elgar, Cheltenham

Fortis M, Carminati M (2012) Il contributo delle «4 A» del made in Italy e dei distretti industriali nella storia dell'export italiano: il caso della meccanica. In: Quadrio Curzio A, Fortis M (eds) L'industria nei 150 anni dell'unità d'Italia Paradigmi e protagonisti. Fondazione Edison Series, Bologna, Il Mulino

Quadrio Curzio A, Fortis M (eds) (2000) Il made in Italy oltre il 2000. Fondazione Edison Series, Bologna, Il Mulino

Quadrio Curzio A, Fortis M (eds) (2007) Valorizzare un'economia forte. L'Italia e il ruolo della sussidiarietà. Fondazione Edison Series. Bologna. Il Mulino

Quadrio Curzio A, Fortis M (eds) (2012) L'industria nei 150 anni dell'unità d'Italia. Paradigmi e protagonisti. Fondazione Edison Series. Bologna. Il Mulino

Vastano S (2013) Il nuovo governo vi salverà. Interview with Martin Schulz by Stefano Vastano. In l'Espresso, 16 May 2013

Chapter 2
Structure and Methodology

Abstract This chapter describes the structure of the essay and explains the research methodology used to estimate the size of the Emilia automatic packaging machinery district and that one of Baden-Württemberg.

Keywords Prodcom classification codes · HS96 international classification codes · Import · Export · Trade balance

Before going into the details of the analysis, the paper traces the evolution of manufacturing in Bologna which saw the extraordinary growth of automatic wrapping and packaging machines, a central pillar of Italian mechanical engineering. It then goes on to describe the history of the Emilia district, highlighting the factors that led to its success.

The analysis proper proceeds with: (a) an examination of the export figures and trade balance for the sector in Italy and Germany based on Eurostat data; (b) ranking of the two countries in global trade taking into account the world's 10 top producing countries of packaging machinery, based on UN data[1]; (c) analysis of the size of the two districts (Emilia and Baden-Württemberg) in terms of number of companies, sales and size of workforce.

(a) *Italian and Germanv exports/imports on the basis of Eurostat data*

For figures on exports and imports of packing and wrapping machines we adopted the Prodcom classification codes used by Eurostat:

- 28292180—Machinery for packing or wrapping (excluding for filling, closing, sealing, capsuling or labelling bottles, cans, boxes, bags or other containers);
- 28292150—Machinery for filling, closing, sealing, capsuling or labelling bottles, cans, boxes, bags or other containers, machinery for aerating beverages.

[1] Data elaborated by Stefano Corradini.

M. Fortis and M. Carminati, *The Automatic Packaging Machinery Sector in Italy and Germany*, SpringerBriefs in Business,
DOI 10.1007/978-3-319-12763-7_2

(b) *Global trade based on UN data*

To rank the two countries in global trade we used the HS96 international classification, and specifically the codes:

- HS 842220—Machinery for cleaning or drying bottles or other containers;
- HS 842230—Machinery for filling, closing, sealing, or labelling bottles, cans, boxes, bags or other containers; machinery for capsuling bottles, jars, tubes and similar containers; machinery for aerating beverages;
- HS 842240—Other packing or wrapping machinery (including heat-shrink wrapping machinery);
- HS 842290—Parts.

(c) *Turnover and workforce of Italian and German companies*

Figures for overall sales and employees for the Emilia district were obtained from the AIDA database for Ateco2007 code:

- 28.29.3—Manufacture of automatic machines for dispensing, packaging and packaging (including parts and accessories).

This data was filled out with balance sheet information on the biggest companies, taken either from Chamber of Commerce statistics or directly from company web sites.

Evaluating the size of the German companies in terms of sales and employees was more complicated owing to the scarcity of available figures. Statistics for the sector compiled by the (VDMA) Verband Deutscher Maschinen und Anlagenbau—German Engineering Federation are only communicated to its members, while the AMADEUS database (that collects business information on Europe's biggest companies) gives data only for broad aggregates of economic activity and does not provide details for individual companies in the specific sector of packaging machinery. We therefore made reasoned estimates on the basis of financial statements for the biggest and best known firms in the sector, taken from AMADEUS.

Chapter 3
Mechanical Engineering in Bologna: Origins and Protagonists. Some Historical Profiles

Abstract This section presents a selection of the principle metalworking firms that emerged in the Bologna area from the dawn of industrialization in the pre-unification era to the present day, in order of foundation.

Keywords Historical profiles · Bologna metalworking industry · Principle metalworking firms · Bologna motorcycle industry

3.1 From the Origins to the Second World War

The history of the Bologna metalworking industry has its origins in the years *prior to Italian unification*, when two private companies in particular dominated the sector:

- CALZONI was founded in 1834 as a repair shop for agricultural machinery and later focused on producing steam engines, machines for oil and grain milling and metal fabrication. In 1907 it was the first Italian company to build hydraulic turbines.
- SOCIETÀ ITALO-SVIZZERA, formerly the "Officina meccanica e fonderia" (founded in Castelmaggiore in 1853 and later transferred to Bologna), produced tractors and steam engines for farming, mainly for large estates (Rinaldi 2008).

We cannot speak about a genuine industrial system until after *Italian unification* when several major companies emerged:

- GAETANO BARBIERI, founded in 1870, initially produced steam boilers and locomotives, machines tools for working iron and wood and other equipment for small manufacturers; it went on to specialize in making refrigeration plants.
- OFFICINE MACCAFERRI, based in Zola Predosa from 1879, produced metal container fencing for river banks and chains to protect premises.
- OFFICINA DELLA SOCIETÀ PER LE STRADE FERRATE MERIDIONALI, which manufactured and maintained rolling stock for the company responsible for building the railways in Southern Italy.

M. Fortis and M. Carminati, *The Automatic Packaging Machinery Sector in Italy and Germany*, SpringerBriefs in Business,
DOI 10.1007/978-3-319-12763-7_3

- LABORATORIO PIROTECNICO DELLO STATO, the state munitions factory founded in 1880 which saw rapid growth during the First World War.
- OFFICINE CEVOLANI, founded in 1900, repaired and built bicycles and produced bottle capsuling machines. In 1934, after the death of the founder Edoardo, CEVOLANI became a precision mechanical and electrical engineering workshop.

Alongside these large companies were a host of small workshops producing a whole range of products from agricultural machinery and surgical instruments to metal gates, furniture and carriages, which later went on to produce machine tools for the paper and food industries (Zamagni 1986; Rinaldi 2008).

So industrialization truly got underway at the end of the 19th century.

A major period of growth in the Bologna metalworking industry occurred from 1903 and 1911 when the number of workers grew dramatically from 2,200 to 9,800, making mechanical engineering the biggest manufacturing segment in the province. During the *Giolittian Era* (1901–1914) demand came above all from local government, the army and the railways. The army commissioned metal racks for meat storage, caging for river defences, bores for artesian wells and machinery for producing ammunition. The railways needed carriages and coaches, and local authorities required mechanical solutions for water supply and equipment for transportation. In this period the main driver of growth in the Bologna machinery segment was therefore the State (Rinaldi 2008).

In 1914 the MASERATI brothers set up their eponymous OFFICINE, initially specializing in repairing automobiles and building racing engines. In 1926 they announced the first wholly Maserati car, the Tipo 26, which won the Targa Florio on its debut[1] driven by Alfieri, one of the Maserati brothers. The most prestigious of the firm's many successes were the consecutive victories in the Indianapolis 500 in 1939 and 1940. In 1937 the Maserati production facilities were sold to the steel manufacturing Orsi family and moved to Modena (Maserati 2012, www.maserati.it; Biblioteca Salaborsa 2012, www.bibliotecasalaborsa.it).

The outbreak of the *First World War* gave a further boost to Bologna mechanical engineering, albeit only briefly. The STABILIMENTO PIROTECNICO STATALE employed 12,000 employees and the military arsenal just over 6,000. CALZONI employed 480 workers and converted to producing machines for the preparation and wrapping of cartridges. OFFICINE BARBIERI had 226 workers in 1916 and began producing refrigerated machines for making explosives. By July 1918 the 104 Bologna mechanical plants which were declared 'auxiliary' to the war effort alone employed over 20,000 people, the vast majority of them women (Rinaldi 2008).

1915 saw the start of SCIPIONE INNOCENTI's career with a small workshop making postage stamps which later converted to producing electromechanical instruments and systems for rail and highway signalling. In 1933 he founded

[1] The Targa Florio is one of the oldest automobile races. Along with the Mille Miglia, it is the most famous Italian race in the world.

SASIB, one of the pioneers of what was to become the world's most important hub for automatic packaging machinery (see Sect. 4.1).

At the end of the Great War many other companies emerged and went on to have great success. Among these we should mention:

- OFFICINE MECCANICHE BOLOGNESI, founded in 1918 and converted in 1921 to SABIEM—Società Anonima Bolognese Industrie Elettromeccaniche, produced 23 different types of machines for a variety of applications, from drawing pasta to making up cigarettes. After 1929 SABIEM specialized in manufacturing electromechanical lifting equipment.
- CARROZZERIA MENARINI & C., founded in 1919 by Ettore Menarini, began with simple bodywork repair and developed by 1925 into a builder of complete industrial vehicles (vans, lorries, ambulances), especially trams and buses, in collaboration with FIAT.
- FABBRICA DI MACCHINE UTENSILI MINGANTI, founded in 1919 by Giuseppe Minganti, designed, produced and sold machine tools (drills, lathes, milling machines, etc.), including spare parts and accessories, also handling installation, maintenance and repair. The presence of Fiat and RIV among its biggest customers was one of the factors behind the success of the company, which became first the main competitor and later the heir to the leading Bologna mechanical engineering firms, from Calzoni to Barbieri.
- OFFICINE di CASARALTA was founded in 1919 by Carlo Regazzoni and Cesare Donati who bought out the munitions firm Sigma after it fell into difficulties after the war. They began with the production and repair of coaches, carriages and mechanical equipment for railways and general purposes, but quickly specialized in building trams.

But the emergence of such firms does not mean that Bologna was immune to the dramatic consequences of post-war conversion. The workforce at the Stabilimento Pirotecnico Statale fell to 2,000 from the 12,000 employed during the war and the Arsenal shed around 6,000 workers.

Despite unfavourable economic conditions for all European manufacturers who were struggling with post-war reconstruction, in the 1920s the Bologna machinery industry found new dynamism and began to express its full potential. This period saw the birth of a number of companies which later grew to considerable size. Among them, in chronological order, were:

- BREVETTI BARONCINI, founded by Ettore Baroncini in 1920, produced special spark plugs for the aviation and automobile industries.
- WEBER, founded in 1923 by Turin engineer and pilot Edoardo Weber. In 1925 it built the first carburettor for the Fiat 501 and grew to occupy a leading position in the European powertrain market eventually reaching a 65 % share. Besides Fiat the company also supplied other leading auto manufacturers including Maserati, Ferrari and Alfa Romeo. After the Second World War it was absorbed by Fiat and subsequently by Magneti Marelli (Biblioteca Salaborsa 2012, www.bibliotecasalaborsa.it).

- ANONIMA COSTRUZIONI MACCHINE AUTOMATICHE—ACMA, founded in 1924 by Gaetano Barbieri, was the outstanding pioneer of the future Bologna packaging machinery district (see Sect. 4.1).
- GD, founded by lawyer Mario Ghirardi and engineer Guido Dall'Oglio in 1923, began as a motorcycle manufacturer. After the war it switched to making packaging machines for sweets, cigarettes and medicines and is today the leading player in the industry (and the district) (see Sect. 4.1). The company played a fundamental role in the growth of the Bologna machinery industry. It represented "not only the coordination point of an articulated and integrated cluster of companies, but also the main nursery for engineers and entrepreneurs who went on to set up the biggest mechanical engineering firms in the province" (Rinaldi 2008). GD spawned many spin-off firms that were together responsible for the extraordinary development of the segment. One of the 1923 founders, engineer Guido DALL'OGLIO, set up by himself to build motorcycles under in his own name. Mario Mazzetti left GD to join Alfonso Morini, Angelo Mattei and Giuseppe Massi-Mazzi and in 1924 they set up another major company, FABBRICA ITALIANA MOTOCICLETTE MORINI—MM which by the outbreak of war was the biggest motorcycle maker in the city. Alfonso Morini in turn later left MM to create MOTO MORINI in 1946. GD also produced Mario Cavedagna, who in 1930 left to found the company CM. Later still, in 1951, Vittorio Minarelli, GD's former chief mechanic, joined Franco Morini (nephew of Alfonso) to create FBM—FABBRICA BOLOGNESE MOTOCICLI[2] manufacturing small cc engines (Rinaldi 2008).
- SOCIETÀ RADIO BREVETTI DUCATI was founded by the Ducati family in 1926 to produce components for the nascent radio broadcasting industry. In the mid 1940s Ducati decided to extend its range of activities and began construction of the Cucciolo (an auxiliary engine for bicycles) acquiring the production rights from its inventor SIATA[3] of Turin, before going on to enjoy huge success in motorbike manufacturing, playing a leading role on the world's motorcycle racing tracks.
- DEMM was founded as a gear maker in Milan in 1919 by the Daldi brothers and Luigi Matteucci; in 1927 it moved to Porretta Terme in the province of Bologna and broadened its output to engine components and precision instruments. Under the technical direction of Giuliano Mazzini, DEMM achieved fame as a manufacturer of mopeds and small-cylinder motorbikes for both touring and racing. In 1956 the DEMM *Siluro* broke 24 world records on the Monza racetrack.
- CURTISA, founded in 1929, produced metal shutters and door/window frames.
- OFFICINA MALAGUTI, set up by Antonio Malaguti in 1930, began as a bicycle manufacturer and in the 1950s switched to building the frames of the celebrated *Mosquito* followed by mopeds and scooters.

[2] FBM went on to produce 20 models a day of the famous Gabbiano, while the 50 cc Pettirosso engine remained in production for 30 years. In 1956 FBM split into two derivations: Franco Morini and Minarelli Motori. By the 1970s the turnover of these two firms far exceeded that of all other Bologna firms in the sector put together. See www.bibliotecasalaborsa.it.

[3] SIATA—Società Italiana Applicazioni Tecniche Auto-Aviatorie was a small Italian automobile manufacturer active from 1926 to 1970.

Bologna therefore established itself from the 1920s as a major centre for motorcycle manufacturing. The traditional local passion for motorcycle racing, which dates back to before the First World War, led to the emergence of numerous bike producers and repair shops. Many other companies from that period, besides the ones above mentioned, specialized in mounting engines on bicycles or assembling components for small cc motorbikes. The models produced by these companies include motorized bicycles like the ABRA (1923), the DIANA (1923–1925) and the MBR (1924–1926), the DE TOGNI side by side two-seater motorbike (1932), and the AQUILA (1926–1935) and AUGUSTA (FIAB) small engine motorcycles.

But the motorbike segment was not the only one that took root and grew in the area. During this period the bases were laid for what would become after the war the pillar of Bologna mechanical engineering, the *automatic wrapping and packaging machines* sector (see Chap. 4). There was also the huge industrial machinery segment, specializing in anything from meat grinders and tomato squeezers (Zocca, Roberti, F.lli Zanardi), to machines for making ice-cream (Cattabriga Otello) or matches (Laffi). Meanwhile agricultural machinery, which CALZONI (initially) and SOCIETÀ ITALO-SVIZZERA had dominated, were on the decline. As we have seen the period also saw closer ties between various Bologna firms (Weber, Minganti, Menarini) and the big manufacturers in the Milan-Turin-Genoa triangle, FIAT especially (Zamagni 1986).

In the decade that followed the most important development was the expansion of DUCATI which in the *1930s* became Bologna's biggest company. The company had still not begun to build the motorcycles which would make it famous worldwide. With the construction of the Borgo Panigale plant in 1935 DUCATI consolidated its leadership in radio and electrical engineering and increased its workforce to 750 and later to 2,000 (Ferretti 2001). The first *Manens* static capacitor was followed by other types (variable in 1928, electrolytic in 1932, and the *Manens* tank capacitor in 1935). From 1936 on, with production of precision radiophonic components well established, DUCATI began to diversify into other devices such as the Dufono, a speakerphone for indoor communications (1935); *Bimar* marine binoculars with night vision, the *Raselet*, the first Italian electric razor (both 1940), and the *Duconta* electronic calculator (1942).

The 1930s also saw the birth of VERLICCHI, initially producing hand-crafted tubular steel parts such as handlebars and exhaust pipes for the motorcycle manufacturers GD, MM and CM. After the Second World War the company began making the complete chassis for mopeds and motorcycles (later the *Cucciolo*, produced by Ducati from 1946, would be mounted on Verlicchi frames) and from the 1950s it supplied Morini, Malanca and Cimatti.

In the *second half of the 1930s* rearmament by the Fascist regime led to a spate of government contracts which drove a new phase of growth for the Bologna metalworking industry. Employment rose dramatically, absorbing huge numbers of women workers from the rural economy. On 16 July, 1935 new legislation decreed that the biggest Bologna manufacturers were to convert to military production. So by 1938 jobs in the industry rose to 22,000 from 12,000 in 1927 (Zamagni 1986; Rinaldi 2008).

Thus the *Second World War* saw a repetition on a larger scale of events during and after the Great War, with the vigorous development of many firms which were required to contribute to the war effort. CALZONI switched to making hydrodynamics systems for aircraft and submarines as well as servo-controls for tanks. SAIBEM began to manufacture artillery components; SASIB made gun-mountings and munitions, and serviced engines for the airforce; ACMA received orders for aircraft torpedoes and engines and for arms construction; BREVETTI BARONCINI went into full-scale production of spark plugs for the airforce, which were regenerated by ALMA; WEBER made carburetors for the German army; OFFICINE CASARALTA made rolling stock; GD built engines for the airforce and the army, as well as parts for Breda anti-aircraft guns; CEVOLANI made weapons parts and aircraft engines; CURTISA switched to making magazines and mines and later pontoon bridges; DUCATI produced radio transmitters and receivers for the army, as well as engine pumps for the Germans, capacitors, optical equipment, weapons parts and aircraft engines; BARBIERI expanded its production of refrigeration and freezer plants; MM had to curtail its output of three-wheel vans and set aside motorcycle production; MINGANTI made milling machines for engine sumps. By the end of 1943 the Bologna firms had converted most of their production to military purposes and, especially during the early years of the conflict, employment grew dramatically. DUCATI employed 7,000 workers (10,000 according to some sources) (Ferretti 2001); the MINGANTI workforce rose from 600 in 1938 to 1,550 in 1941; CALZONI employed 1,600 workers in round-the-clock shifts (Biblioteca Salaborsa 2012, www.bibliotecasalaborsa.it).

In the immediate *post-war period* many firms had to face a serious reconversion crisis, in some cases laying off much of their workforce (Zamagni 1997) [between 1948 and 1954 it is estimated that around 9,000 people were laid off, though some of these sackings were for political reasons (Alaimo and Capecchi 1992)]. The drying up of government contracts led to an inevitable contraction of the biggest local firms, many of which had been seriously damaged by the bombing. But for motorcycle manufacturing they were years of rapid expansion. The motorcycle had become a very different item from previous years, in terms of technical characteristics, look, type of demand satisfied and type of buyer. The costly, fast and heavy pre-war racing bikes were replaced by cheaper, lighter and slower vehicles better suited to city use for the general public (Ferretti 2001).

Among the firms either created or entering the motorcycle sector in this period were (Biblioteca Salaborsa 2012, www.bibliotecasalaborsa.it):

- CIMATTI, founded by cyclist Marco Cimatti in 1937, initially as a bicycle manufacturer, in 1950 entered the 50-cc segment (the so-called "*cinquantini*") producing a vast range of models.
- MALANCA, founded in 1947, began as a maker of bike components and in 1956 began making complete motorcycles.
- MARZOCCHI, founded by the Marzocchi brothers in 1949, specialized in producing telescopic forks and hydraulic shock absorbers which were to be so successful on the Mondial, Ducati and CM bikes.

Beginning in the second half of the 1940s DUCATI (which was saved from collapse by state intervention) saw the potential in motorcycle manufacturing and decided to try and take advantage of it. Without abandoning the radio-electronics sector, in 1946 it began construction of the *Cucciolo*, the first DUCATI motorcycle product. At first sold as a kit, and later with its own chassis, the Cucciolo became the most famous engine of its type in the world, with more than 100,000 units produced. In July 1949 DUCATI began production of its first complete bike (the *Ducati 60*) and 1952 saw the *Cruiser 125*, the first electrical ignition scooter (Ducati 2012, www.ducati.it).

The other pre-war motorcycle manufacturers (except GD which abandoned the industry to devote itself to automatic machinery), returned with traditional bikes updated to meet the new demand, especially smaller cylinder engines.

To sum up, in the first half of century mechanical engineering in Bologna specialized in three main areas:

- radio and film equipment (DUCATI);
- motorcycles (GD, MM, CM);
- machinery, ranging from agricultural machines (CALZONI, DE MORSIER) and machine tools, (BARBIERI, MINGANTI), to *automatic bottling, wrapping and packaging machines* which still today represent the most characteristic segment of Bologna's industry.

The presence in the area of important centres of education undoubtedly played a vital role in driving mechanical engineering in the area in the first half of the century. The university and its faculty of engineering helped to create a robust manufacturing tradition that built upon local manual skills. The Aldini-Valeriani technical school also helped to spread a culture of mechanical engineering, producing many of Bologna's specialist technicians and workers (Alaimo and Capecchi 1992). But the State also played an important part. Major government contracts for munitions and armaments prior to both world wars did much to drive the expansion of Bologna's biggest firms without compromising their versatility or flexibility.

3.2 The 50 Years Following the War and the Economic Miracle

In the *second half of the last century* the industrial structure of the Bologna area saw a reinforcement of existing specializations (Capecchi 1997). From 1951 to 1981 employment in mechanical engineering in Bologna alone quadrupled from 22,000 to more than 86,000. A multitude of small and medium sized firms were set up, predominantly by ex-employees of the older established companies. Rarely did the post-war firms employ more than 500 workers. Sometimes this was a deliberate decision by owners not to risk losing family control. In many cases the owners simply didn't see the need to expand given the presence of a close-knit network of highly qualified and specialized subcontractors for any kind of job (Rinaldi 2008).

In particular the *motorcycle* and *automatic packaging* sectors both had a large number of small and medium sized firms which grew in importance in the 50 years following the war. We will begin by describing the former.

The Bologna motorcycle industry was quick to take advantage of the "economic miracle", offering the public an alternative form of mass transport to the bicycle at a time most working people could still not afford to buy a motor car. From the end of the 1940s to the early 1960s numerous motorcycle makers or related businesses sprang up to join the older established pre-war firms. In 1957 there were around 70, of which only Ducati and Demm had more than 600 employees; all the rest employed fewer than 100 (Rinaldi 2008).

In the 1950s the organizational model of Bologna's motorcycle industry was essentially a more modern version of what had held sway between the wars: a closely integrated network of mostly small and medium sized firms, each one specializing in one or more component. The Marzocchi brothers focused on fabricating forks (from 1952 Marzocchi forks were mounted on Mondial, CM, Maserati and Ducati bikes); Fonderpress focused on welding sumps and engine heads; Morini and Minarelli specialised in engines; Verlicchi in frames; Bonazzi in suspensions, etc. The owners of these firms often came from the older established companies in the area like Ducati, GD, Moto Morini and CM. The companies at the end of the supply chain handled the design, assembly, and machining of some parts and the marketing of the finished product (Ferretti 2001).

In the second half of the 1960s the vitality of the motorcycle sector began to wane, partly due to the disappearance of several well known names, and partly with the end of the intense wave of start-ups that had characterized the industry in previous years. The motorbike began to be replaced by the motor car, now within reach for most Italians and the Bologna motorcycle sector was hit by the recession of the years 1963–1965. In subsequent years production began to expand again, driven by demand for mopeds and scooters which from the mid-Seventies reached the levels of big motorbikes. Meanwhile, however, sales fell off for small-engine motorcycles in which the biggest Bologna firms were specialized. In addition, it was precisely this segment which saw the imperious rise of the Japanese manufacturers, which began to invade foreign markets including Italy starting in the late 1960s. For all these reasons, as well as the financial weakness of many firms, and their scarce attention to marketing and a limited interest in mass production, many small Bologna motorcycle makers disappeared in the 1970s. The companies that survived were those who focused on industrial-scale manufacturing like Ducati or Malaguti, those who moved into segments different from those invaded by the Japanese, such as Malaguti and Italjet, or who despite their high prices continued to hold a fascination for bike enthusiasts, such as Ducati or Moto Morini. Some components makers also managed to become suppliers for some of the world's leading motorcycle manufacturers, such as Minarelli and Franco Morini for engines, or Marzocchi, Grimeca, Verlicchi and Paioli for other components (Ferretti 2001).

References

Alaimo A, Capecchi V (1992) L'industria delle macchine automatiche a Bologna: un caso di specializzazione flessibile. In: D'Attorre PP, Zamagni V (eds) Distretti imprese classe operaia. L'industrializzazione dell'Emilia-Romagna, FrancoAngeli, Milan

Biblioteca Salaborsa (2012) Cronologia di Bologna dall'Unità ad oggi. http://www.bibliotecasalaborsa.it/cronologia/bologna. Accessed 14 Oct 2012

Capecchi V (1997) In search of flexibility: the Bologna metalworking industry, 1900–1992. In: Sabel CF, Zeitlin J (eds) World of possibilities: flexibility and mass production in Western industrialization. Cambridge University Press, Cambridge

Ducati (2012) La storia Ducati. http://www.ducati.it/storia/index.do. Accessed 16 Sept 2012

Ferretti R (2001) Reti di imprese e sistema economico locale. Industria meccanica e comparto motoristico a Bologna (1919–1971). In: Amatori F, Colli A (eds) Comunità di imprese. Sistemi locali in Italia tra Ottocento e Novecento. Il Mulino, Bologna

Maserati (2012) L'azienda. http://www.maserati.it/maserati/it/it/index/maserati/company.html. Accessed 16 Sept 2012

Rinaldi A (2008) I distretti industriali italiani a specializzazione metalmeccanica dalle origini agli anni Novanta. In: Russo M (ed) L'industria meccanica in Italia. Analisi spaziale delle specializzazioni produttive e degli intrecci intersettoriali 1951–2001. Carocci, Rome

Zamagni V (1986) L'economia. In Zangheri R (ed) Bologna. Laterza, Bari

Zamagni V (1997) Una vocazione industriale diffusa. Lo sviluppo industriale in Emilia-Romagna 1880–1990. In: Finzi R (ed) Emilia Romagna, Einaudi, Turin

Chapter 4
The Emilia Automatic Packaging Machinery District: History and Keys to Success

Abstract This chapter describes the emergence and evolution of the Emilia/Bologna wrapping and packaging machinery district, bringing attention to the factors that have made it the success.

Keywords Emilia/Bologna wrapping and packaging machinery district · Origin of the district · Evolution of the district · Factors for the district's success · Anonima Costruzioni Macchine Automatiche (ACMA) · Ghirardi and Dall'Oglio (GD) · Società Anonima Scipione Innocenti Bologna (SASIB) · Aldini-Valeriani Institute

4.1 Origins of the District

As we have seen, wrapping and packaging machinery was the most important sector in post-war mechanical engineering in the Bologna area.

The forefather of the Emilia packaging manufacturing was Anonima Costruzioni Macchine Automatiche (ACMA), founded in Bologna in 1924 by Gaetano Barbieri. The first semi-automatic ACMA machine, the Acma 713, was designed to mechanize the filling of sachets of Idrolitina, a brand of baking soda made by GAZZONI, previously performed by hand. However, orders from Gazzoni (where Barbieri was a partner prior to setting up ACMA) were not enough to guarantee the survival and success of the new company. Nor was the local market, characterized by numerous small manufacturers with only a marginal output of consumer goods. This led ACMA from the outset to address the wider Italian market and focus on the chemicals and pharmaceuticals and confectionery industries with new machines to dose and wrap powders, sweets and chocolates, and print and make paper bags. Meanwhile ACMA's international vocation was quickly in evidence. The first sales in Argentina and Brazil date from the 1920s and the company won an award at the 1929 Barcelona Universal Exposition. By the end of the 1930s its machines were being sold to industrialized economies across Europe and the United States (Alaimo and Capecchi 1992).

M. Fortis and M. Carminati, *The Automatic Packaging Machinery Sector in Italy and Germany*, SpringerBriefs in Business,
DOI 10.1007/978-3-319-12763-7_4

Società Anonima Scipione Innocenti Bologna (SASIB) was founded in 1933 through the transformation of an existing company. In 1937, due to problems in the road and railway signaling sector in which operated, it began to focus on automatic machines in 1937, when it took on an order to build 10 cigarette packaging machines for the Bologna Manifattura Tabacchi (one of the state tobacco manufacturing plants). These machines were designed and patented by a US company, American Machine Foundry (AMF), and the order was awarded to SASIB by the State Monopoly which had acquired the license for their manufacture. Subsequently SASIB began to produce its own cigarette packing machines based on the earlier models, a move which was to prove decisive: unlike the AMF machines, which were covered by patent and could only be sold to the Italian Monopoly, the new machines could be sold anywhere, enabling SASIB to begin to build its own market. In 1942 the manufacture of packaging machines was still in its infancy and development was very gradual, but the first steps had been taken. The sector offered more interesting and lasting prospects than road and railway signaling, thanks especially to new agreements with the American AMF with which SASIB acquired the rights to manufacture all AMF machines and gained access to its sales network. SASIB thus became AMF's European production hub for the tobacco industry. At the same time, besides those produced for AMF, SASIB began to export its own machines which it continued to develop, consolidating its presence in the industry. Although it continued to operate in the signaling sector, this was largely limited to government contracts and by the early 1950s had been far outstripped by sales of automatic machines. In 1957 SASIB was acquired by the U.S. company and Scipione Innocenti left the company he had founded the following year (Alaimo and Capecchi 1992).

GD, founded in 1923 by Ghirardi and Dall'Oglio, began as a small-scale manufacturer of motorcycles, bicycles and spare parts, later making parts for truck engines, especially in the run up to the war when it was declared 'auxiliary' to re-armament. During the war it made parts for machine guns, engines for the airforce and army and the Ursus tank. In the late 1930s GD was bought out by Enzo Seragnoli and at the end of the war, following a suggestion by his cousin Ariosto (who had previously worked in the electrical engineering department at ACMA), he completely transformed the company to produce packaging machines for sweets, cigarette and medicines.

GD right away began to pursue an aggressive strategy. It quickly succeeded in dominating the sweets and confectionary sector which up till then had been the preserve of ACMA. In the Sixties GD expanded into tobacco, competing head on with SASIB, making machines to pack and cellophane-wrap cigarettes (Biblioteca Salaborsa 2012, www.bibliotecasalaborsa.it). Success in this sector was sealed with the appearance of the revolutionary 4350/Pack automatic packaging machine which became the outright leader and led the way for further machines in the coming years which handled the complete tobacco cycle from individual cigarette manufacture to complete packs (Coesia 2012, www.coesia.com).

4.2 The District's Evolution

Up until the Second World War ACMA, SASIB and GD were just three of the many manufacturers rooted in the tradition of small-scale mechanical engineering in the Bologna area dating back to the second half of the 19th century (Alaimo and Capecchi 1992).

At the end of the war numerous engineers and other specialists (designers, electricians, testers, fitters, etc.) left ACMA to found their own companies. Others took their skills and know-how to existing companies. Still others left ACMA to join other firms and then went on to set up on their own. Notable among these are:

- GIOVANNI PRECI, a fitter at ACMA, in 1945 set up a company which initially built machines to make almond brittles and later manufactured packaging machines.
- NATALINO CORAZZA, another fitter at ACMA, started a business repairing automatic machines largely supported by orders from ACMA, and in 1955 began to produce machines to wrap stock cubes, butter and cheese.
- ARIOSTO SERAGNOLI, an electrician at ACMA, instead of founding his own company, joined GD which his cousin Enzo had bought in the late 1930s, converting production entirely from motorcycles to packaging machines, helping to relaunch the firm and go on to become industry leader.
- BRUTO CARPIGIANI, technical director at ACMA until his death (in 1945), in 1942 founded Anonima Riparazioni Costruzioni Automatiche (ARCA), as supplier to ACMA while continuing to run the technical department (shortly before his death he also set up an ice cream machine company under his own-name which became hugely successful).
- ANTONIO MARTELLI, a design engineer at ACMA, in the early 1950s created Costruzioni Automatiche Martelli (CAM) to produce packaging machinery for the pharmaceuticals industry.
- DANTE MANTOVANI, a mechanical engineer, left ACMA to become a key designer at Antonio Martelli's CAM.
- AGOSTINO BILLI set up a firm in 1954 and shortly afterwards began to collaborate with Carle e Montanari, a Milanese manufacturer of machines to make sweets and chocolates, with Billi as technical director of its Bologna unit. The new company later attracted other engineers from ACMA (Alaimo and Capecchi 1992).

In the post-war period the new companies created by technicians leaving ACMA served the demand above all from the food manufacturers and, to a lesser extent, the pharmaceuticals and tobacco industries. But the paper, chemicals and cosmetics sectors soon saw a need to automate product wrapping, both to combat rapidly rising labour costs and also for technical and commercial reasons.

The second larger wave of automatic machine manufacturers began in the 1960s and early 1970s, started by engineers who had themselves left companies founded by former ACMA employees. This second generation differed from the first in that it focused more keenly on pharmaceuticals and cosmetics, and less on food and confectionary, dominated by existing companies.

The second generation of companies included:

- WRAPMATIC, founded in 1960, which produced winding machines for toilet paper and large reams of paper (Alaimo and Capecchi 1992).
- Industria Macchine Automatiche (IMA) was founded in 1961 by ANDREA ROMAGNOLI, a mechanical engineer trained at Aldini-Valeriani who joined ACMA as a design engineer (from 1948 to 1953) and later worked in the GD technical department (from 1953 to 1960) (Rinaldi 2008). Initially the company entered the food packaging segment with a machine for chocolate eggs in 1961. In 1967 it introduced the first machine for making tea bags, a sector in which the company would become world leader. In the 1970s it began to produce blister packaging machines for the pharmaceuticals and cosmetics industries. In 1986 it acquired Farmomac and in 2011 completed the acquisition of Corazza Group (IMA-Industria Macchine Automatiche 2012, www.ima.it).
- FARMOMAC was founded in 1961 by AURELIO RICCI and DINO LULLINI, another technician from Aldini-Valeriani, with a background at ACMA, Carle e Montanari and Zanasi. The company's first venture into the industry was a machine for filling perfume bottles (Alaimo and Capecchi 1992).
- Fabbrica Italiana Macchine Automatiche (FIMA) was founded in 1961 by two former GD engineers (TARTARINI and RABBI). The company's first product was a machine for wrapping Easter eggs. This was followed by machines for wrapping chocolates such as Boeri and Gianduiotti. Alongside food packaging machines they quickly expanded into machines to wrap yarns and thread (Biblioteca Salaborsa 2012, www.bibliotecasalaborsa.it).
- ICA was founded in 1963 by GINO RAPPARINI, formerly with CAM, specializing in dosing machines for granular products.
- Fabbrica Macchine Automatiche Ricci (FAMAR), was founded in 1969 by AURELIO RICCI after his Farmomac experience. The company produced the first fully automatic blister packaging machine thus launching the blister industry. In the 1980s FAMAR was taken over by a large tissue manufacturer and subsequently by a German multinational in the tobacco industry (Famar Packaging 2012, www.famarpackaging.com).

The third generation of companies, created between the 1970s and 1980s, targeted other segments not yet dominated by existing companies. Among these we should mention:

- GRANDI R., created in 1970 by RENATO GRANDI, a former Officine Corazza employee. The company found its own niche market in machines for filling and packaging large sacks of building materials (liquid cement or sand) and agricultural products (fertilizers) (Alaimo and Capecchi 1992).
- AM, founded in 1977 by ANIDRITI and MINELLI, two technicians from FIMA (by this time Nuova Fima), produced machines for packaging rulers and schools rulers.
- MAB, created in 1980 by FRANCO BARILLI, who left IMA to set up his own business producing machines for wrapping (commissioned by IMA it-self) and packaging in boxes.

With the 1980s the growth of new packaging machinery producers around Bologna tapered off, although it retained its competitiveness on world markets. The most significant change in the structure of the packaging industry was the emergence of large groups of independent companies under a single parent, as in the case of GD (today Coesia), IMA, Marchesini and Martelli (Rinaldi 2008).

Between 1981 and 1991 the number of firms and workers in the Bologna area fell, largely in favour of the neighbouring provinces of Modena and Reggio Emilia. This was mainly due to the high density of manufacturing in the town and the lack of suitable growth opportunities for new firms which chose to move out to nearby areas thus expanding the confines of the district.

In 2001 the number of companies and jobs began to grow again both in the province of Bologna itself and across the entire Emilia district. In 2007, prior to the global economic crisis there were 5 fewer companies in the province of Bologna compared with 2001, and 12 fewer in Emilia, reflecting the consolidation of the industry into a smaller number of restructured firms. "The need for continually increasing investments, especially in innovation, has led companies to become bigger and adopt more robust forms of incorporation such as limited companies" (Promo Bologna 2010).

4.3 Factors for the District's Success

As we have seen, the development from a context of a few isolated firms into a proper mechanical engineering industry composed of a hundred companies producing wrapping and packaging machinery only began after the war and consolidated throughout the 1950s (Capecchi 1997; Rinaldi 2008). Employment by GD or earlier by ACMA represented an important source of skills and training for the creation of new entrepreneurs. New firms appeared primarily as spin-offs of the mother company, as technicians and engineers left to set up their own businesses. At least three generations of companies were created in this way, the primary fountainhead of these initiatives being ACMA, as we have seen. Occasionally existing firms decided to convert production, seeking out expert designers of automatic machinery, the most significant case being GD.

The historic reasons for Bologna's success in making automatic packaging machines are many and varied. Among them are the engineering prowess and technical skills that enabled firms to satisfy a broad range of demands for individual machines and complete lines. Alaimo and Capecchi point out the salient characteristics of ACMA's machines—features which were largely maintained by its spin-offs—as vital factors for their success: they were *simple* machines which could be operated by unskilled workers; they were *small*, the biggest being no more than two metres long and often less than a metre; they performed *rapidly* and were extremely *versatile*—machines designed for a specific function could be used to carry out many different jobs. A packaging machine could be used to wrap sweets, tablets, pills or any item with a regular shape. In other cases flexibility was achieved by producing different versions of a basic model, or variations of the same machine: in this case machines had a fixed part and a changeable part, or "format". Often basic models were adapted for individual customers at the trial stage, introducing little changes and adjustments ad hoc. This offering of flexible customized machines capable of rapidly performing complex tasks for different companies was the key to success for the company (Alaimo and Capecchi 1992) and the entire district.

The demand for specific machines tailored to the client's requirements also led to the creation of highly specialized technicians who, besides knowing how to operate the machine, also understood the technical project and were able to design them or at least contribute to their design. This explains how it was possible for people without a higher education to build machines capable of competing internationally. In almost all of these firms the head of the technical department was not an engineering graduate but a mechanic from the Aldini-Valeriani Institute who in most cases did not even have a high school diploma (Alaimo and Capecchi 1992). The presence of the Istituto Aldini-Valeriani in the area played a vital role in the development of mechanical engineering in the area and the packaging machinery sector in particular (Farrell and Lauridsen 2001).

Another crucial factor for the Emilia district's success was the high degree of collaboration between managers, technicians and a broad range of specialist workers, as well as the close collaboration between each company and its network of subcontractors. Companies which served a market often put out a large part of their production to the numerous small-scale local suppliers who were capable of carrying out highly skilled machining work. Relations between the client firm and its subcontractors were so close that sometimes an able worker was encouraged to set up his own business by the parent company itself which guaranteed a loan for the initial purchase of equipment and a stable flow of orders over the long-term (Alaimo and Capecchi 1992).

The foreign literature also recognizes the many strengths that explain the extraordinary success of the Emilia packaging machinery industry, giving particular importance to the role played by the lead firms. According to a study "Interpreting the competition. Italy and Germany states: Behind their success in world export markets for packaging machinery", by the Packaging Machinery Manufacturers Institute: "…The success of the Italian packaging machinery industry in the international marketplace seems to derive from a unique set of circumstances which have come together both as the natural result of Italian geography, culture and industrial organization. And, perhaps most importantly, their success is the result of deliberate improvements taken by lead firms. […] The answer to our question, 'what's the secret?' appears to lie with the following:

- Geographic concentration in the rapidly growing economic 'third' region of Italy (e.d. Emilia Romagna, Veneto, Trentino, Tuscany, Marche).
- Prevalence of micro 'family-owned or founded' firms, exhibiting tightly-bound operations and an unprecedented level of flexibility.
- Emergence of industrial districts or 'clusters' of firms which are not only located near one another but also engage and cooperate in different stages of the production of the same or related products. Industrial districts featured long-standing and informal ties; culture of knowledge and technical know-how being

handed down from generation to generation; expertise in highly-specialized manufacturing; strong entrepreneurial and experimental environment; interaction between academics and industrialists to facilitate the diffusion of design and engineering capabilities; culture of consultation and compromise among the network of small manufacturers—reinforcing the culture for industrial relations based on cooperation and competition and helping to foster a sense of community among local manufacturers.

- Reconstitution of clusters by lead firms which created the best of both worlds through the marriage of management, strategy, and global marketing with the innate strengths of the clusters while keeping the small firm structure intact.
- With lead firms exploring foreign markets close to end-users, the entire business group moves along the innovation learning curve and to adapt more quickly to customer demand.
- The lead firms' presence in a wide variety of international markets provides the firms in the group with the necessary order quantities to achieve economies of scale—hence lower costs leading to better pricing strategies." (PPMI—Packaging Machinery Manufacturers Institute 2004).

References

Alaimo A, Capecchi V (1992) L'industria delle macchine automatiche a Bologna: un caso di specializzazione flessibile. In: D'Attorre PP, Zamagni V (eds) Distretti imprese classe ope-raia. L'industrializzazione dell'Emilia-Romagna, FrancoAngeli, Milan

Biblioteca Salaborsa (2012) Cronologia di Bologna dall'Unità ad oggi. http://www.bibliotecasalaborsa.it/cronologia/bologna. Accessed 14 Oct 2012

Capecchi V (1997) In search of flexibility: the Bologna metalworking industry, 1900–1992. In: Sabel CF, Zeitlin J (eds) World of possibilities: flexibility and mass production in Western industrialization. Cambridge University Press, Cambridge

Coesia (2012) Group history. http://www.coesia.com/en/home/thegroup/grouphistory. Accessed 15 Oct 2012

Famar Packaging (2012) L'esperienza Famar. http://www.famarpackaging.com. Accessed 21 Oct 2012

Farrell H, Lauridsen AL (2001) Collective goods in the local economy: the packaging machinery cluster in Bologna. Mimeo (also in: Crouch C, LeGalès P, Trigilia C, Voelzkow H (eds) Local production systems in Europe, vol II. Oxford University Press, Oxford)

IMA—Industria Machine Automatiche (2012) Group information. http://www.ima.it/. Accessed 23 Oct 2012

PPMI (Packaging Machinery Manufacturers Institute—IDA Consulting) (2004) Interpreting the competition. Italy and Germany: behind their success in world export markets for packaging machinery

Promo Bologna - Agenzia per il Marketing territoriale (2010) Filiera macchine automatiche. Studi e ricerche. December 2010 data

Rinaldi A (2008) I distretti industriali italiani a specializzazione metalmeccanica dalle origini agli anni Novanta. In Russo M (ed) L'industria meccanica in Italia. Analisi spaziale delle specializzazioni produttive e degli intrecci intersettoriali 1951–2001. Carocci, Rome

Chapter 5
Italy and Germany: The Two Leading Countries in the Packaging Machinery Industry

Abstract Italy and Germany are the world's two leading exporters of automatic wrapping and packaging machines. This section presents the situation, beginning with a comparison of the exports and trade balance of the two countries, based on Eurostat data. This is followed by an analysis of the UN figures for the major exporting countries which shows the clear dominance of Italy and Germany in world exports of wrapping and packaging machines.

Keywords Wrapping and packaging machinery industry · German export · Italy export · Geographical breakdown of exports · Territorial concentration

5.1 Exports and the Trade Balance of Italy and Germany According to Eurostat Data: 2008–2012

The Table 5.1 shows figures for the two main segments of the wrapping and packaging machinery industry over the period 2008–2012. It reveals the clear supremacy of Italy, in both exports and trade balance, in the segment "Machinery for packaging or wrapping", and Germany's dominance in the segment "Machinery for filling, closing, sealing, capsuling or labelling".

The data show how both countries were hit by the crisis that began in 2008, leading to a dramatic falloff in exports, but also that the damage was less severe in Italy than in Germany. In 2009 Italy saw its exports of "Machinery for packaging or wrapping" fall by −17.9 % against the previous year compared with a decline of 36.4 % over the same period for German exports of "Machinery for filling, closing, sealing, capsuling or labelling". Taking an aggregate of the two segments German exports fell by −26.6 % while Italian exports fell by −19.4 %.

In addition, while Italian exports in both segments began to grow again as early as 2010, German exports continued to contract, only recovering in 2011. But even more important is the fact that by 2012 Italy had already fully recovered and exceeded its pre-crisis levels, while Germany remained well below this threshold. German exports for its top segment stood at 2.6 billion euros in 2008, fell to 1.5 in 2010 and

M. Fortis and M. Carminati, *The Automatic Packaging Machinery Sector in Italy and Germany*, SpringerBriefs in Business,
DOI 10.1007/978-3-319-12763-7_5

Table 5.1 Value of export and trade balance of German and Italian packaging machinery: 2008–2012 (million euros)

	Export					Trade balance				
	2008	2009	2010	2011	2012	2008	2009	2010	2011	2012
Cod. 28292180—Machinery for packing or wrapping										
Germany	1,584	1,418	1,380	1,711	1,768	1,315	1,189	1,150	1,433	1,481
Italy	1,726	1,416	1,555	1,911	1,987	1,600	1,331	1,448	1,802	1,872
Cod. 28292150—Machinery for filling, closing, sealing, capsuling or labelling										
Germany	2,594	1,650	1,514	1,571	1,897	2,481	1,545	1,410	1,426	1,730
Italy	1,124	881	1,096	1,165	1,234	1,029	793	973	1,058	1,116
Cod. 28292180 + Cod. 28292150										
Germany	4,178	3,068	2,894	3,282	3,665	3,795	2,734	2,560	2,859	3,211
Italy	2,850	2,298	2,651	3,076	3,221	2,629	2,124	2,421	2,861	2,988

Source Compiled by Fondazione Edison on Eurostat Database (Eurostat 2014)

rose to 1.9 billion in 2012. Meanwhile Italian exports in its top segment were 1.7 billion euros in 2008, fell to 1.4 in 2009 and rose to 2 billion in 2012. The aggregate data show 4.2 billion euros for German exports in 2008 and 3.7 billion euros in 2012, a full half a billion below its pre-crisis peak. Italian exports stood at 2.8 billion and reached 3.2 billion by 2012, around 400 million euros more than 2008.

The same goes for the trade balance where Italy is ahead in "Machinery and equipment for packing or wrapping" and Germany prevails in "Machinery and equipment for filling, closing, sealing or labelling". Again by this measure Italy in 2012 amply exceeded its pre-crisis levels, both overall and taking the two segments separately. Germany, instead, remained below this level both in its leadership segment and taking the two segments together.

To sum up the historic comparison, over the 5 years period (2008–2012) the gap between Italy and Germany narrowed steadily both in exports (from 1.3 billion euros in 2008 to 0.4 billion euros in 2012), and in terms of the trade balance (from 1.2 to 0.2 billion euros).

The latest data for the period November 2012–October 2013 (Fig. 5.1), confirm the marked narrowing in the gap between German and Italian exports of packaging machinery compared with pre-crisis peaks. In the period November 2012–October 2013 the gap was 0.4 billion euros, against a gap of 1.4 billion in the period October 2007–September 2008. The figure clearly illustrates trends in Italian and German exports of automatic machinery according to the Standard International Trade Classification (SITC). The segment in question is SITC 74527 (Other packing/wrapping machinery).

The figure also reveals that while German exports are still well below pre-crisis levels, Italian exports have not only made up ground following the collapse of global trade, but have abundantly exceeded the previous 2008 peak for Italian exports in the industry.

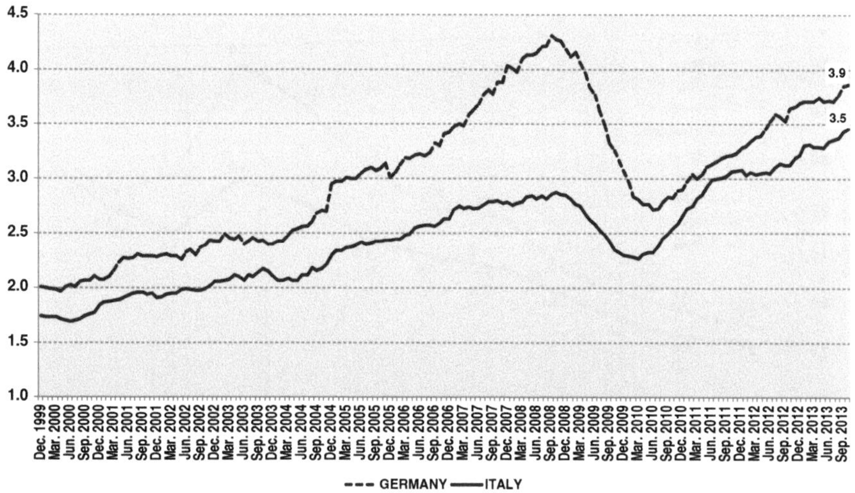

Fig. 5.1 Export of packaging machinery, code SITC 74527 (billion euros, "sliding" past 12 months, updated to October 2013). *Source* Compiled by Fondazione Edison on Eurostat Database (Eurostat 2014)

5.2 UN Data on Italian and German Exports: 2000–2011

The importance of Italy and Germany in global exports of packaging machinery is also confirmed by the UN data. The Fig. 5.2 clearly shows how their historical international dominance in the sector is growing. The figure is based on aggregate exports of products for the 4 HS96 international classification codes for automatic wrapping and packaging machines [see Chap. 2, (b)].

German exports of wrapping and packaging machinery rose from 2.5 billion dollars in 2000 to 5.7 billion in 2009, after peaking at 8 billion dollars in 2008. Following the collapse in 2009 (−2.3 billion against 2008) in 2010 German exports saw a further decline to 5.3 billion dollars. Only in 2011 did they return to growth, reaching 6.2 billion dollars.

Meanwhile Italian exports grew from 2.2 billion dollars in 2000 to 4.5 billion dollars in 2009, peaking at 5.6 billion dollars in 2008. Thus the decline in 2009 was only 1.1 billion dollars against 2008, and already by 2010 Italian exports of packaging machines stood at 5 billion dollars, well on the way to reaching pre-crisis levels. In 2011 there was a further marked rise in Italian exports to 6 billion dollars, well above pre-crisis levels, unlike German exports which remained well below the peak.

By 2011 the German exports' lead over Italian exports had thus shrunk considerably, to just 0.2 billion dollars.

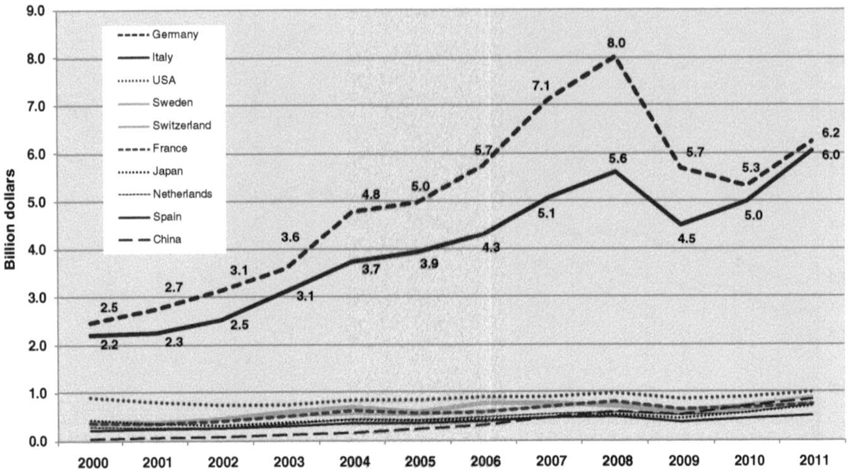

Fig. 5.2 Time series of world exports of the top 10 producing countries in the packaging machinery industry (aggregate date derived from the sum of codes 842220, 842230, 842240, 842290 of the HS96 international classification): 2000–2011. *Source* Compiled by Fondazione Edison on UN Comtrade Database (UN Comtrade 2013)

5.2.1 Analysis by Segment

1. "Other packing or wrapping machinery, including heat-shrink wrapping machinery"

Italy holds the leadership in exports of "Other packing or wrapping machinery, including heat-shrink wrapping machinery" (Fig. 5.3).

This is the biggest segment in the automatic wrapping and packaging machines industry. In 2011 total exports came to 8.7 billion dollars, or 39.2 % of exports for the sector as a whole.

After the collapse of 2009, Italian exports saw an initial recovery as early as 2010 (from 2 to 2.1 billion dollars), while German exports declined further from 2 to 1.8 billion dollars.

In 2011 Italian exports continued their strong recovery to reach 2.7 billion. Germany also returned to growth, reaching 2.4 billion. Both exceeded their pre-crisis peaks, though Italy slightly rather more markedly.

2. "Machinery for filling, closing, sealing or labelling bottles, cans, boxes, bags or other containers; machinery for capsuling bottles, jars, tubes and similar containers; machinery for aerating beverages"

Germany historically holds the leadership for exports of "Machinery for filling, closing, sealing or labelling bottles, cans, boxes, bags or other containers; machinery for capsuling bottles, jars, tubes and similar containers; machinery for aerating beverages" (Fig. 5.4). This is the second biggest segment in the industry. In 2011 total exports amounted to 7.1 billion dollars, or 31.8 % of exports for the entire packaging machinery industry.

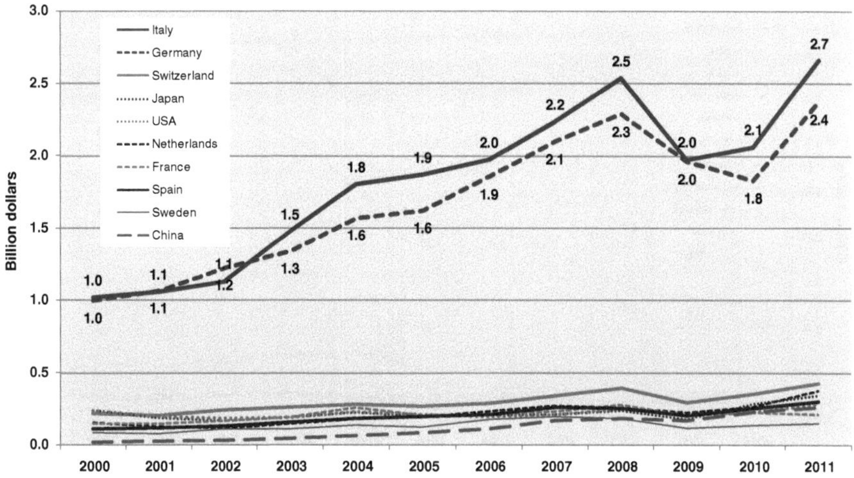

Fig. 5.3 World exports of the top 10 producing countries in the sector "Other packing or wrapping machinery (including heat-shrink wrapping machinery)": 2000–2011. *Source* Compiled by Fondazione Edison on UN Comtrade Database (UN Comtrade 2013)

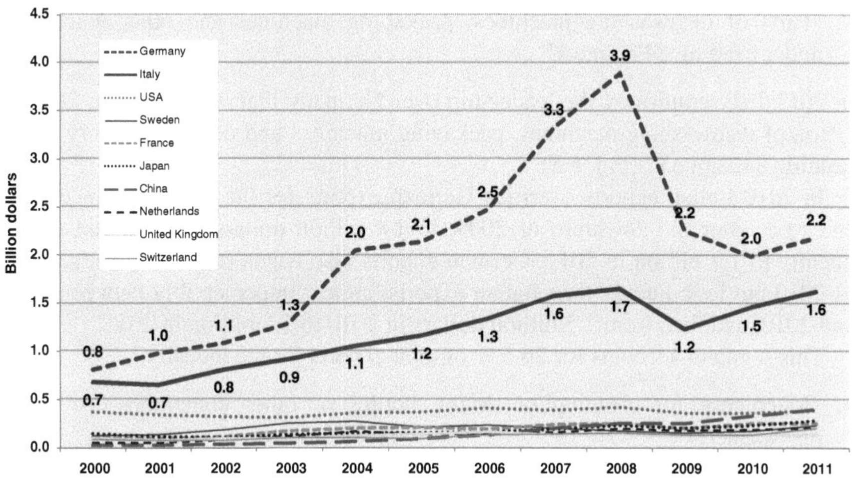

Fig. 5.4 World exports of the top 10 producing countries in the sector "Machinery for filling, closing, sealing, or labelling bottles, cans, boxes, bags or other containers; machinery for capsuling bottles, jars, tubes and similar containers; machinery for aerating beverages": 2000–2011. *Source* Compiled by Fondazione Edison on UN Comtrade Database (UN Comtrade 2013)

After the peak of 2008 and the subsequent collapse of 2009, in 2010 German exports saw a further sharp decline, from 2.2 billion dollars in 2009 to less than 2 billion dollars in 2010, before rising again to 2.2 billion in 2011. Italian exports instead began to gain ground as early as 2010 rising from 1.2 billion dollars in 2009 to around 1.5 billion dollars in 2010 and 1.6 billion in 2011, perceptibly closing the gap with Germany.

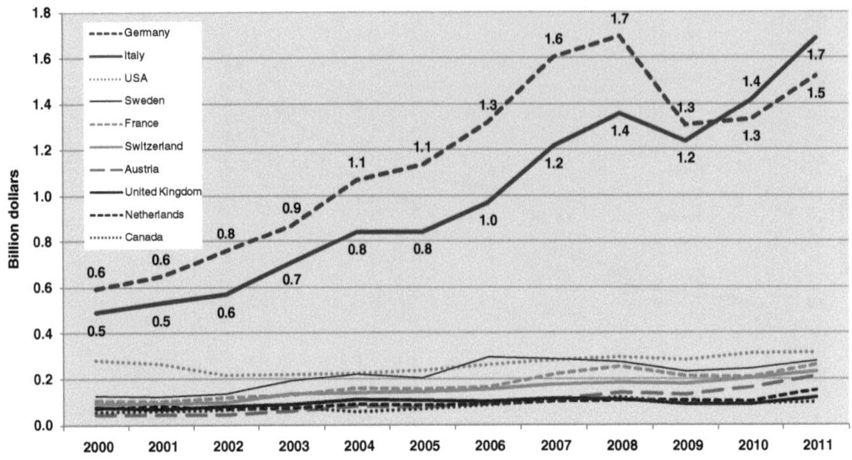

Fig. 5.5 World exports of the top 10 producing countries in the sector "Parts of dishwashing machines, machines for the packing and other machinery of code 8422, nes": 2000–2011. *Source* Compiled by Fondazione Edison on UN Comtrade Database (UN Comtrade 2013)

3. "Parts of dishwashing machines, packaging machines and other machinery under heading 8422, n.e.s."

In 2011 Italy confirmed the leadership over Germany that it had won in 2010 in "Parts of dishwashing machines, packaging machines and other machinery under heading 8422, n.e.s" (Fig. 5.5).

In 2010 Italian exports overtook German exports for the first time since 2000 and rose, after the downturn of 2009, to 1.4 billion dollars in 2010 and subsequently to 1.7 billion in 2011. German exports also began to grow again as early as 2010 but less quickly than Italian exports, almost imperceptibly between 2009 and 2010, and then from 1.3 billion dollars in 2010 to 1.5 billion in 2011.

This segment accounts for 26.5 % of total exports for the industry.

4. "Machinery for cleaning or drying bottles or other containers (excluding dishwashers)"

Finally the export leader for "Machinery for cleaning or drying bottles or other containers (excluding dishwashers)" is Germany (Fig. 5.6). However, this is a very small segment. In 2011 total exports amounted to just 550 million dollars, or 2.5 % of exports for the entire packaging machinery industry. German exports in the same year amounted to 164 million dollars.

Italy, with exports worth 75 million dollars in 2011, contested second place with Sweden which in 2011 reported exports of 87 million dollars, confirming its position as the world's second exporter in this tiny segment.

However in 2011 Sweden's exports fell slightly against 2010, (from 88 to 87 million dollars), while those of Italy (from 66 to 75) and Germany rose (from 155 to 164).

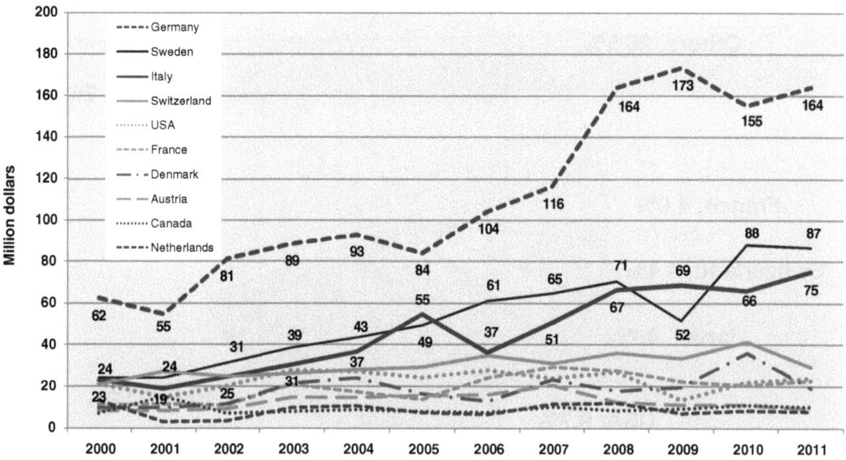

Fig. 5.6 World exports of the top 10 producing countries in the sector "Machinery for cleaning or drying bottles or other containers": 2000–2011. *Source* Compiled by Fondazione Edison on UN Comtrade Database (UN Comtrade 2013)

5.3 Italy and Germany: The Dominating Forces of the World Market

Italy and Germany together dominate the sector of automatic wrapping and packaging machines. According to UCIMA Research department estimates (UCIMA-Italian Packaging Machinery Manufacturers Association 2014) Italy and Germany hosting the 36 % of World production in packaging machinery industry, that has been around 30.5 billion euros in 2012.

This chapter examines in detail the geographical breakdown of world exports and the main export destinations for Italy and Germany.

5.3.1 *Geographical Breakdown of Exports*

The Figs. 5.7 and 5.8 give two snapshots of the breakdown of exports world-wide for the years 2000 and 2011.

They show that in 2000 just over 50 % of world exports in the industry came from Germany and Italy, the former slightly ahead (26.7 %) of the latter (24 %). In 2011 the combined quota of the two countries rose to over 55 %, again with a slight weighting in favour of Germany (28 %), but with Italy making up ground (27.1 %).

In particular:

- Germany's share of world exports rose from 26.7 % in 2000 to 28 % in 2011, an increase of 1.3 points.
- Italy's share rose from 24.0 % in 2000 to 27.1 % in 2011, a rise of 3.1 points.

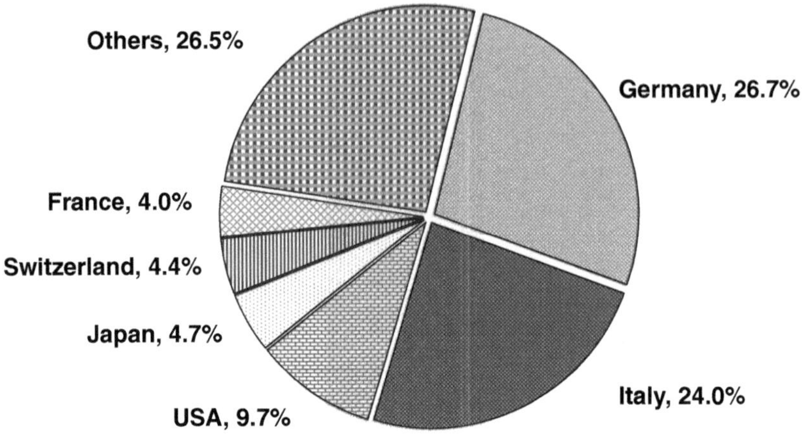

Fig. 5.7 World exports in the packaging machinery industry (aggregate data derived from the sum of codes 842220, 842230, 842240, 842290 of the HS96 international classification) (year 2000). *Source* Compiled by Fondazione Edison on UN Comtrade Database (UN Comtrade 2013)

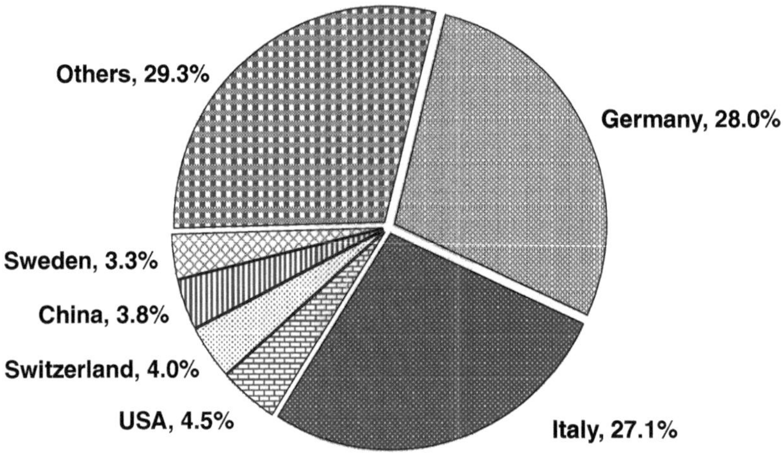

Fig. 5.8 World exports in the packaging machinery industry (aggregate data derived from the sum of codes 842220, 842230, 842240, 842290 of the HS96 international classification) (year 2011). *Source* Compiled by Fondazione Edison on UN Comtrade Database (UN Comtrade 2013)

In 2011 the gap between Germany and Italy was less than 1 % point, compared with a gap of 2.7 % points in 2000.

5.3.2 Main Export Destinations

The Figs. 5.9 and 5.10 show the main exports markets for Italian and German packaging machines.

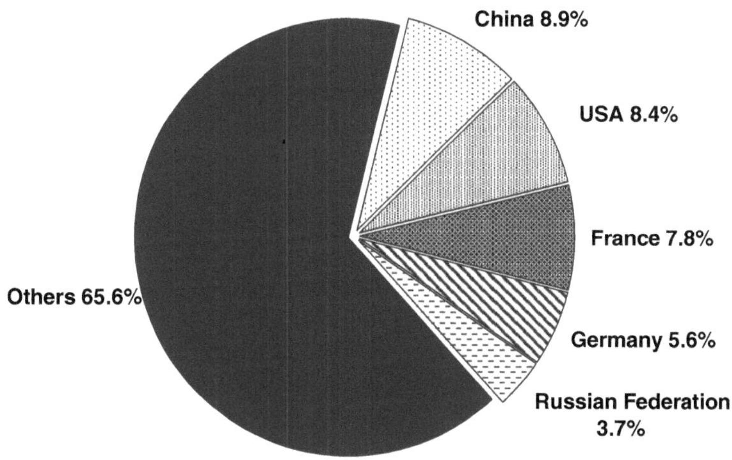

Fig. 5.9 Italy: main countries of destination of exports of packaging machinery (aggregate data derived from the sum of codes 842220, 842230, 842240, 842290 of the HS96 international classification) (year 2011). *Source* Compiled by Fondazione Edison on UN Comtrade Database (UN Comtrade 2013)

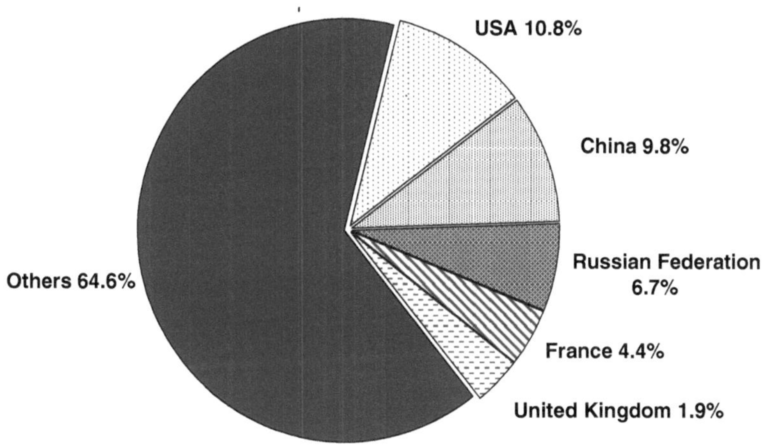

Fig. 5.10 Germany: main countries of destination of exports of packaging machinery (aggregate data derived from the sum of codes 842220, 842230, 842240, 842290 of the HS96 international classification) (year 2011). *Source* Compiled by Fondazione Edison on UN Comtrade Database (UN Comtrade 2013)

As we see in 2011 the biggest outlet for Italy's exports in the sector was China (8.9 %), followed by the United States (8.4 %); France (7.8 %), Germany (5.6 %) and Russia (3.7 %).

Germany exported mainly to the United States (10.8 %), China (9.8 %), Russia (6.7 %), France (4.4 %) and the U.K. (1.9 %).

Germany appears among the biggest importers of Italian goods (5.6 %), while Italy does not figure among the main destinations for German exports in the sector.

5.3.3 Territorial Concentration

Besides the fact that Germany and Italy both hold positions of clear international leadership in wrapping and packaging machines, they also share a geographical concentration of production in well-defined and highly specialized territorial hubs.

In Italy the main concentration is in Emilia Romagna, known as "Packaging Valley". Lombardy, Piemonte and Veneto also have their own manufacturing centres, but Bologna is the real capital of packaging machinery. Together the companies in Emilia Romagna, Lombardy, Piemonte and Veneto represent more than 80 % of the firms in the industry, but Emilia Romagna alone has around 50 % of them and 70 % of Italy's sector turnover (Promo Bologna 2010a, b). Therefore the greatest concentration of the industry is in Emilia Romagna: not only in Italy, but, as we shall see, in the world.

In Germany the biggest cluster is in the state of Baden-Württemberg to the north-east of Stuttgart, in the contiguous districts of Schwäbisch Hall and Waiblingen. Other clusters can be found in North Rhine-Westphalia and Lower Saxony, but these are relatively minor hubs. The districts of Schwäbisch Hall and Waiblingen are the two areas that dominate German manufacture of packaging machinery.

Both Schwäbisch Hall and Waiblingen have recently acquired institutional status, the first through the creation of Packaging Valley (PV), the second through the Packaging Excellence Center (PEC), two associations which bring together many—though not all—of the major industry players.

References

Eurostat (2014) Eurostat statistics. http://epp.eurostat.ec.europa.eu/portal/page/portal/statistics/. Accessed 8 Jan 2014

Promo Bologna-Agenzia per il Marketing territoriale (2010a) Filiera macchine automatiche, Studi e Ricerche. December 2010 data

Promo Bologna-Agenzia per il marketing territoriale (2010b) Bologna, the "Packaging Valley". Studi e Ricerche.

UCIMA-Italian Packaging Machinery Manufacturers Association (2014) UCIMA Research Department. Financial analysis of budgetary data

UN Comtrade (2013) UN Comtrade database. http://comtrade.un.org/. Accessed 15 Jan 2013

Chapter 6
The Size of the Emilia Packaging Machinery District

Abstract In this chapter we have calculated the estimated dimensions of the Emilia packaging district both on the basis of a more extended geographical area including the provinces of Reggio Emilia and Parma, and a more restricted area covering only the two core provinces of Bologna and Modena.

Keywords Emilia packaging machinery district · Bologna · COESIA · IMA · SACMI · MARCHESINI

As we have seen, the biggest Italian manufacturing hub for wrapping and packaging machines is in Emilia Romagna (Fig. 6.1). The heart of the district is the province of Bologna, but it extends as far as the provinces of Modena, Reggio Emilia and Parma. Taking the 4 provinces of Emilia together the size of the Italian district is similar to that of the German district of Baden-Württemberg, stretching around 100 km.

However we have calculated the estimated dimensions of the Emilia packaging district both on the basis of a more extended geographical area including the provinces of Reggio Emilia and Parma, and a more restricted area covering only the two core provinces of Bologna and Modena (Tables 6.1 and 6.2).

The metrics considered are the number of companies, employees and turnover, as shown in the two tables below, for ATECO 2007 (NACE) code 28.29.30: "Manufacture of automatic machines for dispensing, packaging and packaging (including parts and accessories)."

As regards the number of companies, our analysis only covers those companies for which balance sheet data is available for the years in question. For the international groups (Coesia, Marchesini and IMA Spa) the figures for employees and revenues are taken from their respective consolidated statements. For Sacmi Group, which operates in many different businesses, we took the figure for employees and revenues from the divisions involved directly in the packaging industry (Beverages & Packaging, Plastic Division, Food Machinery and Inspection). For the province of Modena we excluded from code 28.29.30 the figures for Tetra Pack Solution Spa which is part of the Swedish Tetra Pak group. For the province of Parma we excluded the figures for Corima International Machinery Srl, which is part of the Marchesini group.

M. Fortis and M. Carminati, *The Automatic Packaging Machinery Sector in Italy and Germany*, SpringerBriefs in Business,
DOI 10.1007/978-3-319-12763-7_6

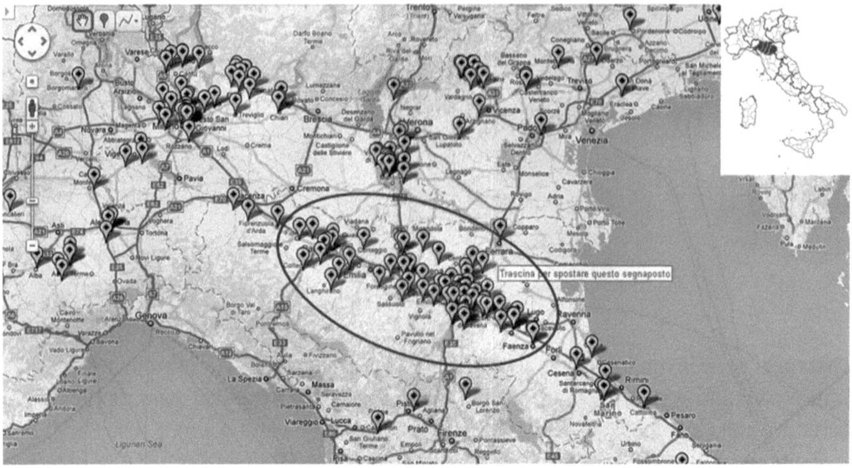

Fig. 6.1 Territorial concentration of Italian companies in the sector of packaging machinery. *Source* Compiled by Fondazione Edison on AIDA Database (Bureau van Dijk 2012)

Table 6.1 The size of the Emilia packaging machinery district in the 'more extended' geographical version

Province	2011			2010		
	No. of enterprises	Employees	Turnover 000 euro	No. of enterprises	Employees	Turnover 000 euro
Bologna	117	13,143	2,732,984	118	10,604	2,331,040
Modena	26	754	160,869	25	332	98,656
Parma	29	2,047	812,157	30	1,723	728,324
Reggio Emilia	12	180	66,693	12	116	54,507
Total	184	16,124	3,772,703	185	12,775	3,212,527

NACE code 282930—Manufacture of automatic machines for dispensing, packaging and packaging (including parts and accessories)

The number of firms only refers to companies for whom balance sheet data for that year is available

In the case of international corporations (Coesia—Marchesini—Ima Spa) figures for the number of employees and turnover are taken from their respective consolidated financial statements. In the case of Sacmi the data used to calculated the number of employees and turnover only refer to the divisions directly involved in the packaging industry (Packaging, Plastics and Food Divisions)

For the province of Modena, Tetra Pack Solution Spa (part of the Swedish Tetra Pak Group) is not included. In 2010 data is not yet included for Gruppo Fabbri Vignola Spa constituted 24 November, 2010 (the single turnover figure for 2011 amounted to EUR 44.8 million with 268 employees)

For the province of Parma, Corima International Machinery Srl (part of Marchesini Group) is not included

Source Compiled by Fondazione Edison on AIDA Database (Bureau van Dijk 2012) and Mediobanca data (Mediobanca 2012)

Table 6.2 The size of the Emilia packaging machinery district in the 'more restricted' geographical version

	2011			2010		
	No. of enterprises	Employees	Turnover 000 euro	No. of enterprises	Employees	Turnover 000 euro
Bologna	117	13,143	2,732,984	118	10,604	2,331,040
Modena	26	754	160,869	25	332	98,656
Total	143	13,897	2,893,853	143	10,936	2,429,696

NACE code 282930—Manufacture of automatic machines for dispensing, packaging and packaging (including parts and accessories)
The number of firms only refers to companies for whom balance sheet data for that year is available
In the case of international corporations (Coesia—Marchesini—Ima Spa) figures for the number of employees and turnover are taken from their respective consolidated financial statements. In the case of Sacmi the data used to calculated the number of employees and turnover only refer to the divisions directly involved in the packaging industry (Packaging, Plastics and Food Divisions)
For the province of Modena, Tetra Pack Solution Spa (part of the Swedish Tetra Pak Group) is not included
In 2010 data is not yet included for Gruppo Fabbri Vignola Spa constituted 24 November, 2010 (the single turnover figure for 2011 amounted to EUR 44.8 million with 268 employees)
Source Compiled by Fondazione Edison on AIDA Database (Bureau van Dijk 2012) and Mediobanca data (Mediobanca 2012)

Taking all 4 Emilia provinces together there were 184 'pure' manufacturers of packaging machines in 2011, 117 of which were in the province of Bologna, employing over 16,000 workers, with revenues of more than 3.7 billion euros, around 72 % generated by firms in Bologna province (Table 6.1).

If we look instead at the provinces of Bologna and Modena alone, in 2011 there were 143 packaging machine firms, with nearly 14,000 workers and total revenues of around 2.9 billion euros (Table 6.2).

So even using this more restricted definition, the figures for the Emilia packaging machinery district are quite considerable.

Finally here is a brief round-up of the major packaging machine manufacturers in the Emilia district—COESIA, IMA, SACMI and MARCHESINI—all located in the province of Bologna (Table 6.3).

The COESIA-Seragnoli group is the "product" of several evolutions of GD (see Sect. 4.1), founded in 1923 to build motorcycles, and bought out by Enzo Seragnoli in the late 1930s. It grew rapidly in the 1940s and 1950s in sweets and soap wrapping and in the 1960s became market leader in the tobacco sector with the revolutionary 4350/Pack. Coesia is the named adopted in 2005 to identify the holding as a group founded on cohesion and shared goals and values. The group is controlled entirely by Isabella Seragnoli, and made up of 12 companies: in the last 10 years acquisitions of HAPA, LAETUS, NORDEN, CITUS-KALIX, ADMV, SACMO, SASIB and FLEXLINK have been added to CIMA, ACMA, GDM and VOLPAK which it acquired since the 1980s as part of a strategy of product diversification. The group presently has 60 operating units (48 of which are production

Table 6.3 Major manufacturers of packaging machinery in the Emilia district

Industrial group	Province	2012		2011	
		Employees	Turnover 000 euro	Employees	Turnover 000 euro
GRUPPO COESIA (*) (Fam. Seragnoli—Isabella Seragnoli)	Bologna	5,500	1,200,000	5,000	973,000
GRUPPO IMA	Bologna	3,561	734,310	3,529	669,210
Gruppo SACMI (**)	Bologna	1,704	515,000	1,690	489,000
GRUPPO MARCHESINI	Bologna	1,000	204,000	1,000	187,000
Total		11,765	2,653,310	11,219	2,318,210

(*) The 2012 turnover data is proforma and includes the acquisition of RA Jones: 1.36 billion of euros

(**) The data used only refers to the divisions involved in the packaging industry (Packaging, Plastic and Food Divisions)

Source Compiled by Fondazione Edison on AIDA Database (Bureau van Dijk 2012)

plants) in 27 countries. In 2012 revenues amounted to 1.2 billion euros and the headcount was 5,500, making Coesia today the biggest Italian manufacturer of automatic machinery (Coesia 2012, www.coesia.com).

Industria Macchine Automatiche (IMA) was founded in Bologna in 1961. In 1963 a 52 % stake was bought by the Vacchi family which turned it from a Società in Accomandita Semplice (limited partnership), into a Società per Azioni (joint-stock company) and encouraged the company's growth. From being a small local firm IMA soon became an international group, specializing initially in machines for making tea bags and later diversifying production. Today Gruppo IMA is world leader in the design and production of machines for the processing and wrapping of tea and coffee, as well as pharmaceuticals, cosmetics and foodstuffs in general. The group is structured around two main brands: IMA Industries and IMA Pharma. The first is world leader in the design and production of machines for the packaging of tea, coffee and beverages, and for the processing and wrapping of food products, cosmetics and toiletries. It include six companies and divisions: TEA & COFFEE DIVISION, GIMA S.p.A., CORAZZA S.p.A., STEPHAN MACHINERY GmbH, BFB DIVISION, REVISIONI INDUSTRIALI S.r.l. The second is world leader in machines for the processing and packaging of pharmaceuticals, and contains three highly specialized divisions: IMA ACTIVE (Solid Dose Solutions), IMA LIFE (Aseptic Processing & Freeze Drying Solutions), IMA SAFE (Packaging Solutions). Today the group has 22 production facilities in Italy, Germany, the U.K., the United States, India and China with a sales network operating in more than 70 countries (IMA 2012, www.ima.it). 2012 revenues amounted to 734 million euros with more than 3,500 employees.

Società Anonima Cooperativa Meccanici Imola (SACMI) was set up in 1919 in Imola as a cooperative of 9 mechanics and blacksmiths led by Romeo Galli and

Giulio Miceti. In the early years business was very limited and the members suffered violence and intimidation by the regime. Despite this the cooperative survived and in the 1930s made its first own-name machine for peeling oranges. After the Second World War growth picked up, when it began to produce presses for tile manufacturing. Cooperativa Ceramiche Imola was one of the firms worst hit by wartime bombing and the Sacmi mechanics repaired the tile presses which had been badly damaged. Gradually Sacmi came to build all the machines (including the furnaces) necessary for tile manufacture, covering the entire ceramics manufacturing cycle. Meanwhile, immediately after the war SACMI began to build machines to produce crown cork bottle tops and later on containers mainly for food manufacturers (Bassani 1999). Today SACMI is an international group and world leader in machines for ceramics, packaging (including beverages and closures & containers), food and plastics. Figures for the divisions directly involved in the packaging sector (Beverages & Packaging, Plastic Division, Food Machinery and Inspection) give revenues in 2012 of 515 million euros and more than 1,700 employees (out of a total workforce of around 3,500).

Gruppo MARCHESINI was created in 1974 at Pianoro, in the province of Bologna, when Massimo Marchesini decided to go into business after making the first packaging machine in his garage. Since then the group, which is still wholly owned by the Marchesini family, has steadily expanded through acquisitions and partnerships with complementary companies and today includes 14 production divisions, 7 acquired companies and 1 partner firm, all specialized in the design and production of different types of packaging machines. Around 85 % of company production is aimed at the pharmaceuticals industry, while the remaining 15 % is in the cosmetics sector. Over 85 % of revenues come from exports, mainly to Europe and the United States (Marchesini Group 2012, www.marchesini.com). In 2012 revenues came to 204 million euros, with more than 1,000 employees.

As can be seen from the Table 6.3, in 2012 these 4 firms had a combined turnover of 2.7 billion euros. In 2011 total revenues came to 2.3 billion euros representing around 60 % of turnover for the entire ‘extended’ Emilia district and more than 80 % of turnover for the more “restricted” district. They therefore represent the pillar of the entire Emilia district, around which revolve numerous medium-small firms.

References

Bassani A (1999) C’erano una volta nove meccanici… Ottant’anni di crescita SACMI, La Mandragora, Imola Bureau van Dijk - Company information and business intelligence (2012), AIDA Database

Bureau van Dijk-Company information and business intelligence (2012), AIDA Database

Coesia Group (2012) Il gruppo. http://www.coesia.com/it/home/thegroup/Factsfigure. Accessed 23 Oct 2012

IMA (2012) Il gruppo. http://www.ima.it/Page/IT/Il_Gruppo-P435.html. Accessed 23 Oct 2012

Marchesini Group (2012) Press kit. http://www.marchesini.com/news-events/press-kit/. Accessed 23 Oct 2012

Mediobanca (2012) Le principali società italiane

Chapter 7
The Size of the German Packaging Machinery District

Abstract In this chapter we have calculated the estimated dimensions of the packaging districts of Schwäbisch Hall and Waiblingen in Baden-Württemberg.

Keywords Schwäbisch Hall/Waiblingen packaging machinery districts · Baden-Württemberg · German Packaging Valley · Packaging Excellence Center

7.1 The Main Areas of German Manufacturing in Automatic Wrapping and Packaging Machines

To identify the main areas of German production we referred to several sources.

7.1.1 The Ivo Mossig Study

The first source is a study done in 2001 by Ivo Mossig of Bremen University, entitled "Evolutionary development of regional production clusters. A case study of the packaging machinery industry in Germany", updated in 2011 together with Lars Schieber. The Fig. 7.1 shows how the Baden-Württemberg district is one of the biggest areas of employment in Germany.

The state of Baden-Württemberg actually has two contiguous districts specialized in packaging machinery: the first is in the region of Schwäbisch Hall, the second is in Waiblingen, both to the east of Stuttgart. These two districts extend for around 100 km, analogous to the more extended Emilia district covering the four provinces of Bologna, Modena, Reggio Emilia and Parma. The two German districts, as we have seen, have acquired institutional status, with the creation of Packaging Valley (PV) and the Packaging Excellence Center (PEC) (Fig. 7.2).

M. Fortis and M. Carminati, *The Automatic Packaging Machinery Sector in Italy and Germany*, SpringerBriefs in Business,
DOI 10.1007/978-3-319-12763-7_7

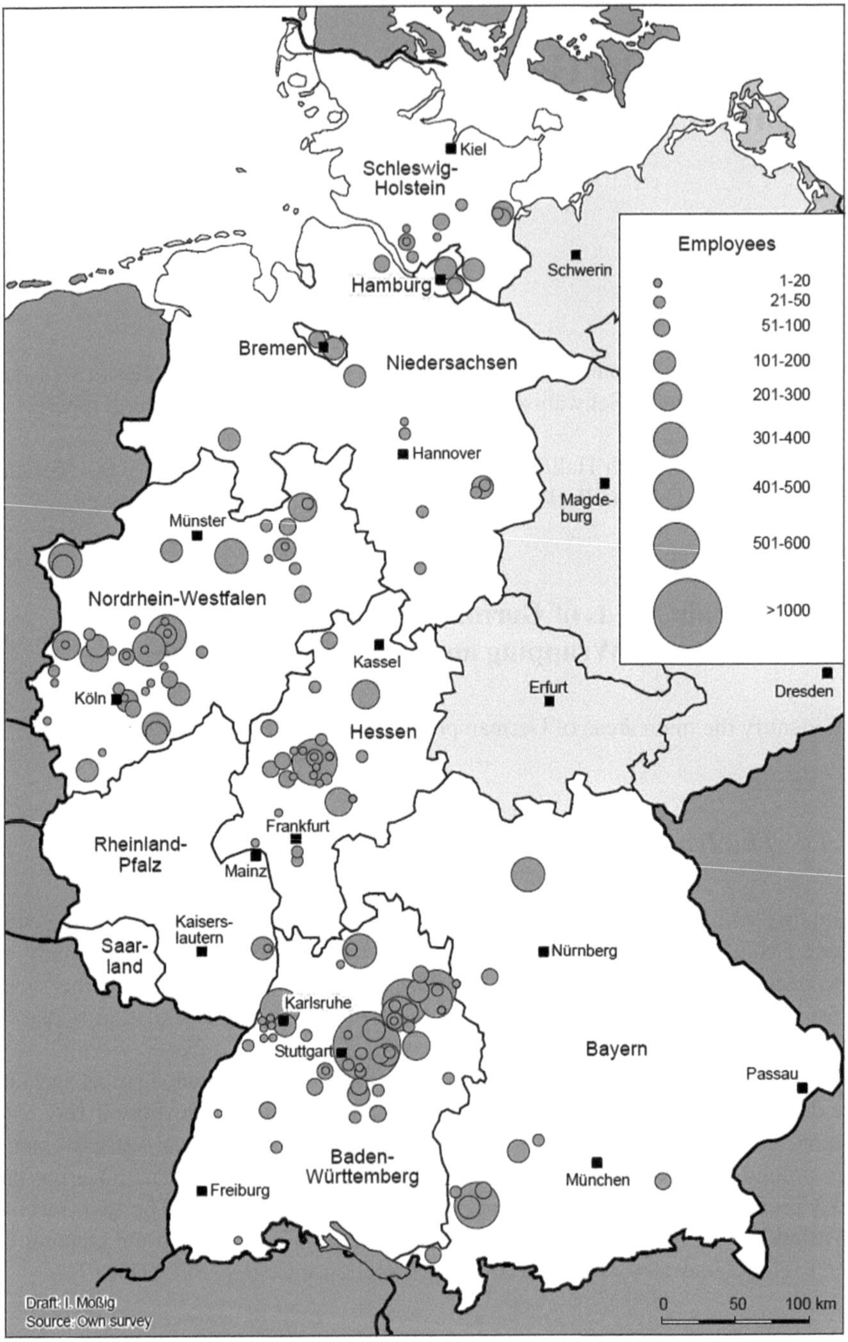

Fig. 7.1 The spatial distribution of packaging machinery factories according to numbers of employees in 1999. *Source* Mossig (2001)

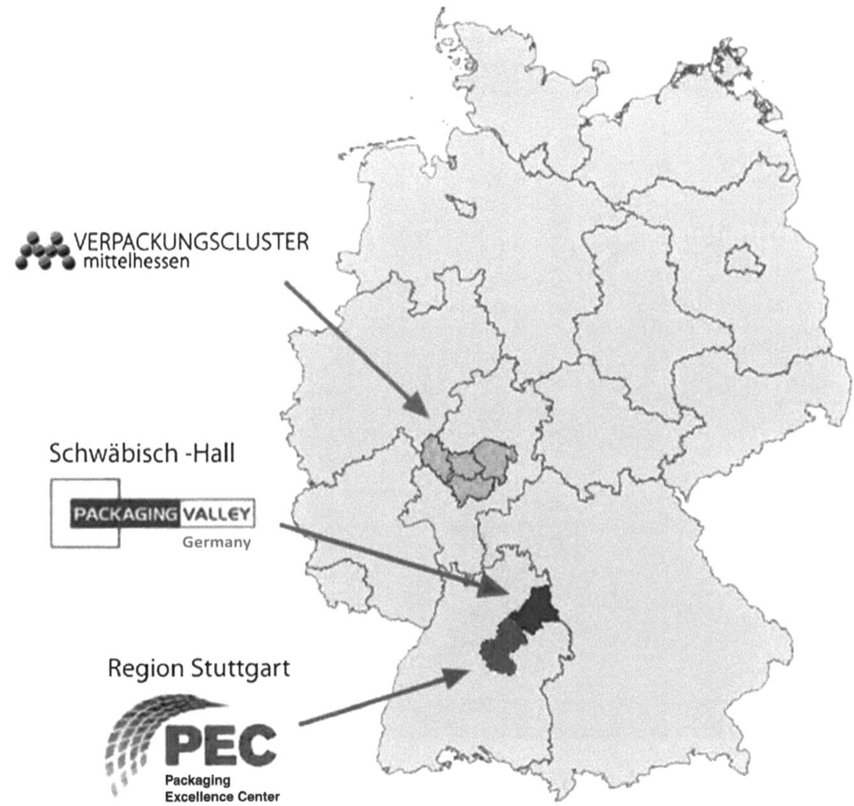

Fig. 7.2 The major regions and district initiatives of the packaging machinery industry. *Source* Schieber and Mossig (2011)

7.1.2 The Fondazione Edison Study

The point of departure for the Fondazione Edison study was an official publication marking 150 years of the German packaging industry, entitled: *Inventors. Doers. World market leaders. 150 Years of Packaging Machinery in Southwest Germany*. In the appendix the publication lists all firms belonging to the Packaging Excellence Center (PEC) and Packaging Valley (PV), giving the number of employees, year of incorporation, location and a brief description of manufacturing activity for each member. We filled these lists out with companies not belonging to the two associations, but named on the VDMA web site under the heading "Packaging Machines" (VDMA 2007, 2011a, b), and plotted their location on a map to examine the territorial concentration (Fig. 7.3). As we can see the Fondazione Edison map corresponds fairly closely with the Ivo Mossig and Lars Schieber study, revealing the concentration of firms to the North-east of Stuttgart, in the districts of Schwäbisch Hall and Waiblingen in Baden-Württemberg.

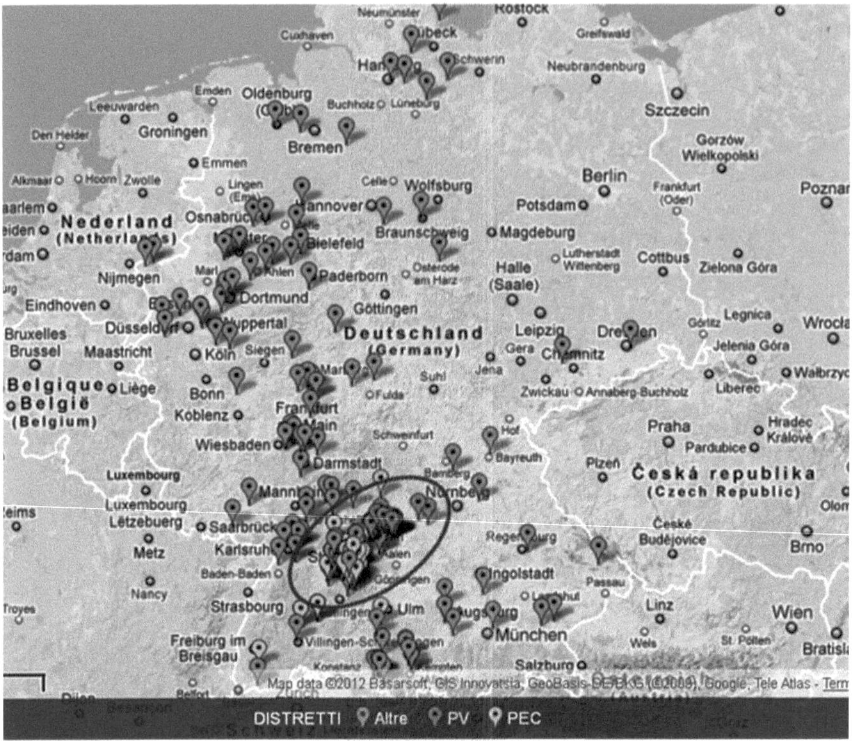

Fig. 7.3 The analysis of Fondazione Edison. *Source* Compiled by Fondazione Edison on Packaging Valley (2011) and VDMA (2007, 2011a, b)

7.2 How Big Are the Two Adjoining Districts of Schwäbisch Hall and Waiblingen? Difficult to Say!

It is far from easy to estimate the size of the two packaging machinery districts in Baden-Württemberg since information is very scarce. One estimate, albeit approximate, of the size of the two districts in terms of jobs, comes from Packaging Valley (PV), the association of the main packaging manufacturers in the district of Schwäbisch Hall, and from Packaging Excellence Center (PEC), the association of firms in the district of Waiblingen.

According to the Packaging Valley web site the combined workforce of the two districts Schwäbisch Hall–Waiblingen is 15,000 (www.Packaging-Valley.com). Of these 15,000 around 7,000 work for 40 packaging machinery firms belonging to Packaging Valley, created in 2007. This is confirmed by the Packaging Valley web site, and the 2011 publication celebrating the Baden-Württemberg packaging machine industry *"Inventors. Doers. World market leaders. 150 Years of Packaging Machinery in Southwest Germany"*.

With reference to 2011, there is a discrepancy between the number of associated companies indicated on the web site (40) and the figure given in the publication (50) at page 202. In addition, the appendix listing the firms belonging to Packaging Valley (and the Packaging Excellence Center) names only 35 companies, plus the 3 biggest firms in the industry (Bausch and Ströbel; Gerhard Schubert; Groninger), for a total 38 companies. This is different again from the figure of 40 or 50.

Fondazione Edison's research also finds discrepancies in the number of workers in the firms belonging to Packaging Valley. Summing up the number of employees indicated in the appendix we arrive at fewer than the total given in the publication itself or on the web site: 4,020 employees against the 7,000 indicated. And even adding in the workers at the 3 biggest companies (Bausch and Ströbel; Gerhard Schubert; Groninger) in the Packaging Valley region (Schwäbisch Hall), which are not members of the association, still gives us a lower figure of 6,312.

Furthermore, some of the member companies are not pure packaging machinery producers, but essentially suppliers of services and industrial solutions. Counting only the employees of the packaging machinery companies, the number of employees fall to 5,025, as can be seen in the Table 7.1, where service and solution providers are shown in light grey.

To find the number of actual packaging machine manufacturers, and thus estimate the size the Schwäbisch Hall district, we look once again at the publication "*Inventors. Doers. World market leaders. 150 Years of Packaging Machinery in Southwest Germany*". At pages 58–59 the book provides a family tree of all wrapping and packaging machine companies belonging to the Packaging Valley and the Packaging Excellence Center. Each firm has either a continuous or dotted line box depending on whether it is a packaging machinery manufacturer or a service provider. Our calculations take into account the first but not the second (shown in light grey in the Table 7.1).

A different number of employees again for the district of Schwäbisch Hall (Table 7.2) appears in the Ivo Mossig study updated with Lars Schieber in 2011: Clusterentwick-lung und politik in der Verpackungsmaschinenbau-Industrie Baden-Württembergs, which gives the Packaging Valley workforce as 5,900.

Let us now turn to the question of the *number of workers employed by the packaging industry in Waiblingen*. As we have seen the region is represented by the Packaging Excellence Center (PEC), an association set up in 2007 that brings together around 50 companies in the packaging and automation sector.

The number of employees in PEC companies is not specified anywhere, but we can arrive at a figure as the difference between the total given on the Packaging Valley web site (which gives the total workforce for PV and PEC as 15,000) and the number of employees in PV which they state is 7,000. Thus the number of employees in the district of Waiblingen should be around 8,000.

Here again it is not possible to simply sum up the number of workers at the PEC firms listed in the appendix of the book on German packaging manufacturing, as the core business of these companies is often automation in general and only marginally related to packaging machinery. In addition, as with Packaging

Table 7.1 List of all member companies of the German packaging valley (PV)

	Companies	Locality	Employees
1	Aichele Werkzeuge GmbH	74564 Crailsheim	54
2	Binder GmbH	74564 Crailsheim	38
3	BPS Packaging Solutions Willy Bader GmbH	74542 Braunsbach	3
4	BREITNER Abfüllanlagen GmbH	74523 Schwäbisch Hall	95
5	E + K Sortiersysteme GmbH	74592 Kirchberg/Jagst	66
6	ETW Wollmershäuser GmbH	74549 Wolpertshausen	23
7	GASTI Verpackungsmaschinen GmbH	74523 Schwäbisch Hall	86
8	GEPAS mbH	74549 Wolpertshausen	12
9	HEITEC AG	74564 Crailsheim	500
10	Hohenloher Verpackungs-Systeme GmbH	74549 Wolpertshausen	20
11	HO-MA GmbH	74532 Ilshofen	9
12	Horlacher + Hintsche Packaging Technology	74523 Schwäbisch Hall	8
13	INOPLAN-Ingenieurbüro	74523 Schwäbisch Hall	3
14	Ishida GmbH	74523 Schwäbisch Hall	26
15	Itek GmbH	74523 Schwäbisch Hall	15
16	ITQ GmbH	74523 Schwäbisch Hall	30
17	Kachel Technische Dokumentation und Kommunikation	74523 Schwäbisch Hall-Sulzdorf	21
18	Kocher-plastik Maschinenbau GmbH	74429 Sulzbach-Laufen	420
19	OCS Checkweighers GmbH	74523 Schwäbisch Hall	55
20	OPTIMA packaging group GmbH	74523 Schwäbisch Hall	1,200
21	PCA Roboter- und Verpackungstechnik GmbH	91578 Leutershausen	60
22	Popp Maschinenbau GmbH	74564 Crailsheim	20
23	PS mako GmbH packaging systems & machines	74544 Michelbach/Bilz	40
24	R. WEISS Verpackungstechnik GmbH & Co. KG	74564 Crailsheim	450
25	Rudolf Deckert GmbH & Co. KG	74523 Schwäbisch Hall	40
26	S. F. Vision GmbH	74523 Schwäbisch Hall	45
27	Seidenader Automation GmbH & Co. KG	74523 Schwäbisch Hall	360
28	ServoTech GmbH	74595 Langenburg	35
29	So-matec epple GmbH	74429 Sulzbach/Laufen	12
30	SorTec Pharma Sortiertechnik GmbH	74549 Wolpertshausen	25
31	Stako GmbH	74549 Wolpertshausen	8
32	Stern Fördertechnik GmbH	74523 Schwäbisch Hall	6
33	SVZ Maschinenbau GmbH	74564 Crailsheim	12
34	Transnova-RUF Verpackungs- und Palettiertechnik	91522 Ansbach	200
35	VMS-Maschinenbau GmbH	74423 Obersontheim	23
	Employees for members of PV		**4,020**
36	Bausch + Ströbel	74532 Ilshofen	950
37	Gerhard Schubert Gmbh	74564 Crailsheim	620
38	Groninger & co. Gmbh	74564 Crailsheim	722
	Employees for the region of PV		**6,312**
	Employees for members of PV excluding providers of industrial solutions		**5,025**

NB Companies providing services or industrial solutions shown in light grey
Source Elaboration of the authors on data from Packaging Valley (2011)

Valley, many companies provide packaging industry services and solutions and some of them have their headquarters outside Baden-Württemberg.

As a result not all the associate members of PEC were taken into account but only those who effectively and primarily produce packaging machinery located

Table 7.2 Evolution of employment in the German Packaging Machinery Industry

	Germany	Schwäbisch-Hall
Employees 1998	25,000	3,500
Employees 2010	27,500	5,900
% change	10	69

Source Schieber and Mossig (2011)

in the state of Baden-Württemberg; service providers or generic automation companies (e.g. Rexroth which has 35,000 employees or Rockwell Automation which has 20,000) were therefore ignored. Apart from these more striking cases, to decide which firms to include we referred to the book *Inventors. Doers. World market leaders. 150 Years of Packaging Machinery in Southwest Germany*, which we used to estimate the employees of the Schwäbisch Hall district.

Stripped of Rexroth and Rockwell Automation, the Table 7.3 looks as follows, with suppliers of services or industrial solutions shown in light grey.

The Fondazione Edison calculation of the workforce in pure packaging machinery manufacturers belonging to PEC is 7,800. This is slightly lower but close to the 8,000 obtained as the difference between the figure for the combined Schwäbisch Hall-Waiblingen districts and the figure for Schwäbisch Hall alone taken from the Packaging Valley web site.

Therefore the Fondazione Edison figure for the total number of employees for packaging machinery producers in Packaging Valley and the Packaging Excellence Center is 12,825.

Finally, to complete the picture, the Table 7.4 shows a list of other manufacturers of packaging machinery in the Schwäbisch Hall and Waiblingen regions but not members of either Packaging Valley or Packaging Excellence Center, whose number of employees is unknown.

We can assume however that these firms are quite small since they do not appear in the map with the major companies given in Ivo Mossig's study (Fig. 7.4).

Although based on 1998 data, the above figure roughly describes a situation that corresponds to the present. The companies named in 1998 as having the biggest number of employees ($\geq$300), are also the biggest in 2010. These are Bosch Packaging System, Optima, Gerhard Schubert, Bausch and Ströbel, Groninger and Kocher-plastik. To these we should add Weiss which, from fewer than 300 employees in 1998, had grown in 2010 to employ 450.

The figures for the major manufacturers are shown in the Table 7.5.

Bosch Packaging Technology is one of the leading manufacturers in the packaging industry. Present in more than 15 countries worldwide, it develops and offers complete manufacturing solutions for the pharmaceuticals, food and confectionary sectors. 2010 revenues for the Packaging Technology division came to 717 million euros, with more than 4,400 employees.

Optima, founded in 1922, is today one of the biggest German producers of automatic wrapping and packaging machines. The target industries are food, cosmetics, chemicals, pharmaceuticals and non-tissue products for baby and adult

Table 7.3 List of all member companies of the Packaging Excellence Center (PEC)

	Companies	Locality	Employees
1	Alfred Giesser Messerfabrik	71364 Winnenden-Birkmannsweiler	50
2	AMK Arnold Müller Gmbh & Co. KG	73230 Kirchheim/Teck	750
3	Balluff GmbH	73765 Neuhausen a.d.F.	2,180
4	BBF Sterilisationsservice GmbH	71394 Kernen-Rommelshausen	23
5	beck packautomaten GmbH & Co. KG	72636 Frickenhausen	85
6	Beckhoff Automation GmbH	72336 Balingen	1,220
7	Bernd Kußmaul GmbH	71384 Weinstadt-Großheppach	42
8	Bosch Packaging Systems GmbH	73630 Remshalden	4,471
9	Bosch Rexroth AG	70736 Fellbach	n.d.
10	Bräuninger Technologie	71384 Weinstadt	16
11	Catalent Germany Holding 2 GmbH	73614 Schorndorf	500
12	Contexo GmbH	73650 Winterbach	40
13	ESR-Systemtechnik Gmbh	70734 Fellbach	9
14	F+P CAD-Systeme	71384 Weinstadt	6
15	FANUC Robotics Deutschland GmbH	73765 Neuhausen a.d.F.	5,000
16	FIX Maschinenbau GmbH	71404 Korb	120
17	Harro Höfliger Verpackungsmaschinen GmbH	71573 Allmersbach im Tal	500
18	HPF Hans P. Friedrich Elektronik GmbH	73660 Urbach	15
19	Klingele Papierwerke GmbH & Co. KG	73630 Remshalden	1,000
20	KOMET Maschinenfabrik GmbH	73207 Plochingen	25
21	KÖRA-PACKMAT Maschinenbau GmbH	78667 Villingendorf	90
22	Leuze electronic GmbH + Co. KG	73277 Owen	720
23	Löffler Verpackungstechnik GmbH & Co. KG	71034 Böblingen	5
24	LTK Lineartechnik Korb GmbH	71332 Waiblingen	35
25	Murrelektronik GmbH	71570 Oppenweiler	1,300
26	Pilz GmbH & Co. KG	73760 Ostfildern	1,300
27	PR-PACK Verpackungstechnik GmbH	71336 Waiblingen	5
28	Rockwell Automation GmbH	76227 Karlsruhe	n.d.
29	ROTHWEIN Technik Sondermaschinenbau	71332 Waiblingen	0
30	Rybak + Höschele GmbH + Co.KG	71332 Waiblingen	35
31	S. Tränkner GmbH Sondermaschinen	71332 Waiblingen	15
32	Schreyer Sondermaschinen GmbH	71549 Auenwald	24
33	SCHUNK GmbH & Co. KG Spann- und Greiftechnik	74348 Lauffen/Neckar	1,800
34	Siemens AG	70049 Stuttgart	1,700
35	SONOTRONIC	76307 Karlsbad	200
36	sortimat Assembly Technology	71364 Winnenden	500
37	STAMBERA VERPACKUNGSMASCHINEN GMBH	70736 Fellbach	25
38	Uhlmann Pac-Systeme GmbH & Co. KG	88471 Laupheim	1,300
39	Widmann Maschinen GmbH	73278 Schlierbach	100
40	WITTENSTEIN alpha GmbH	97999 Igersheim	1,300
41	ZAHORANSKY AG	79674 Todtnau-Geschwend	600
	Total companies joining the PEC		**27,106**
	Number of companies joining the PEC excluding companies providing industrial solutions		**7,800**

NB Companies providing services or industrial solutions shown in light grey
Source Elaboration of the authors on data from Packaging Valley (2011)

Table 7.4 List of other producers of packaging machinery in the district, but not members of PEC or PV

	Companies	Locality
1	Albert Fezer, Maschinefabrik	73730 Esslingen
2	AVT Automatisierte Verpackungs - Technologie Gmbh	71732 Tamm
3	Burghardt Verpackungsmaschinen GmbH	70176 Stuttgart
4	CTA GmbH Verpackungsservice	74321 Bietigheim-Bissingen
5	DUO PLAST AG	36341 Lauterbach
6	EHRET Control GmbH	71570 Oppenweiler
7	Fakura Fahrzeuge Und Fahrzeugteile-Handel Gesellschaft Mit Beschrankter Haftung	70839 Gerlingen
8	Geset Etikettiersysteme Gmbh	73734 Esslingen
9	Hugo Beck Maschinenbau GmbH & Co. KG	72581 Dettingen an der Erms
10	Irex Etiketten und Auszeichnungsgeräte GmbH & Co. KG	72649 Wolfschlugen
11	IVECO GmbH & Co. KG	70839 Gerlingen
12	Kallfass Verpackungsmaschinen GmbH	72622 Nürtingen
13	Medicon GmbH	74523 Schwäbisch Hall
10	Paal VT Verpackungs Technik GmbH & Co. KG	73630 Remshalden
11	Schneider-Kennzeichnung GmbH	70374 Stuttgart
12	Söhnel Maschinenbau, Dieter Söhnel	73262 Reichenbach
13	Strapex GmbH	71088 Holzgerlingen
14	Technologiezentrum Schwäbisch Hall GmbH	74523 Schwäbisch Hall
15	Würschum GmbH	73760 Ostfildern

Source Elaboration of the authors on data from www.maceuro.com and www.industrystock.it

personal hygiene. In 2010 the company had revenues of 225 million euros and employed 1,200 workers.

Gerhard Schubert develops packaging solutions and machinery. Its target sectors are pharmaceuticals, cosmetics, food and beverages, confectionary, frozen products and dairy products. In 1984 it introduced the first robots for packaging. In 2010 it had a turnover of 113 million euros, with 734 workers.

Bausch+Ströbel, founded more than 40 years ago by 4 people, designs, builds and markets packaging and production systems for the pharmaceutical, cosmetics and chemicals industries. Today it is one of the leading players in pharmaceuticals packaging. 2010 revenues came to 105 million euros with 900 employees.

Groninger & Co. was founded in 1980 and produces packaging systems for chemicals, pharmaceuticals and cosmetics. With 514 employees in 2010 it reported a turnover of 65 million euros.

Created in 1963, Kocher-plastik is specialized in packaging sterile liquids and ointments. The head count in 2009 (the last available figure) was 390, with revenues of 58 million euros.

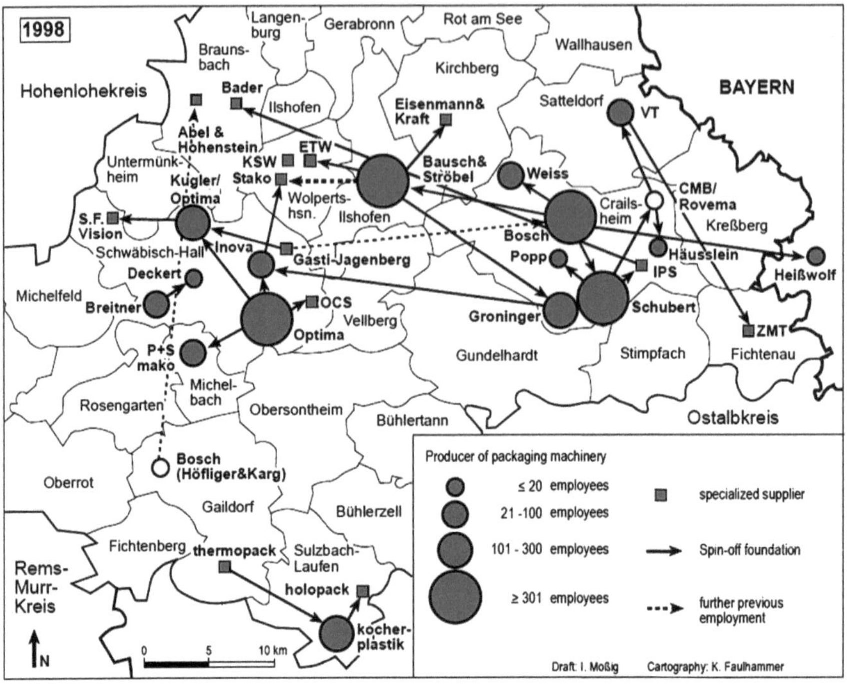

Fig. 7.4 Spin-off foundations in the administrative district of Schwäbisch-Hall 1998 (ca. 70 km northeast of Stuttgart). *Source* Mossig (2001)

Table 7.5 Major manufacturers of packaging machinery in the district of Baden-Wuerttemberg (Schwäbisch Hall and Waiblingen) (Baden-Wuerttemberg)

Industrial groups	2010		2009	
	Turnover 000 euro	Employees 2010	Turnover 000 euro	Employees 2009
Bausch + Ströbel Maschinenfabrik Ilshofen GmbH + Co. KG	105,000	900	105,000	760
Bosch GmbH, Robert, Packaging Technology	717,000	4,471	703,000	4,450
Gerhard Schubert Gesellschaft mit beschränkter Haftung	112,825	734	123,216	733
Groninger & Co. GmbH	65,479	514	62,045	510
OPTIMA-Maschinenfabrik Dr. Bühler GmbH & Co. KG	225,973	1,211	205,848	1,103
kocher-plastik Maschinenbau-GmbH	n.d	n.d	57,581	390
R.WEISS Verpackungstechnik Gmbh & Co. KG	n.d.	450	n.d.	n.d.
Total	1,226,277	8,280	1,256,691	7,946

Source Compiled by Fondazione Edison on AMADEUS Database (Bureau van Dijk 2012) and Bosch data (Bosch 2011)

Finally, R. Weiss, founded in 1990, offers solutions for the packaging of pharmaceuticals, cosmetics and chemicals, foodstuffs, drinks and tobacco, sweets, snacks and other non-food products. In 2010 it had 450 employees.

Altogether, in 2010 these 7 companies had revenues of 1.2 billion euros, employing around 8,300 workers.

References

Bosch—Bosch Packaging Technology (2011) Packed as promises—150 years of packaging technology. In: Interpack 2011—Press Conference

Bureau van Dijk-Company information and business intelligence (2012), AMADEUS Database

Mossig I (2001) Evolutionary development of regional production clusters. A case study of the packaging machinery industry in Germany. Geographisches Institut der Justus-Liebig-Universität, Gießen

Packaging Valley (2011) Inventors. Doers. World market leaders. 150 years of packaging machinery in Southwest Germany

Schieber L, Mossig I (2011) Clusterentwicklung und-politik in der Verpackungsmaschinenbau-Industrie Baden-Württembergs. Beiträge zur Wirtschaftsgeographie und Regionalentwicklung Nr. 1-2011, Universität Bremen, Institut für Geographie

VDMA (2007) World market for packaging machinery 2007 supply according to countries

VDMA (2011a) German Packaging Machinery Industry back to growth path. VDMA news

VDMA (2011b) Germany is the worldwide largest confectionary machinery manufacture, VDMA news

Chapter 8
Emilia and Schwäbisch Hall-Waiblingen: The Two Districts Compared

Abstract The chapter concludes with a summing up of the size of the packaging machinery Italian district and the German Schwäbisch Hall/Waiblingen districts, in terms of employees and turnover.

Keywords Size of the packaging machinery Italian district · Size of the packaging machinery German district · Employees · Turnover

We conclude with a summing up of the sizes of the two districts.

We have seen that the Emilia packaging machinery district, taken to include the provinces of Bologna, Modena, Parma and Reggio Emilia, is very similar in geographical extension to the German Schwäbisch Hall/Waiblingen districts, both covering around 100 km.

The comparison is done on 2010 figures (that for Italy are lacklustre compared to 2011) as at the moment the update to 2011 is not available for the German firms.

In terms of employees and turnover the two districts compare as follows.

(a) **Comparison in terms of employees**

In terms of employees the biggest district is Emilia, which in 2010 had just over 13,000 employees, against the figure of 12,825 calculated for the German district.

Even taking into account the smaller firms which do not belong to PV and PEC, we think the German district may at most equal the size of the Emilia district, but not exceed it.

Among the major groups it appears that all 4 of the biggest Italian companies have a thousand or more employees, while only 2 German firms are of this size.

(b) **Comparison in terms of revenues**

We can only compare sales for the biggest companies in each district because they are the only German firms for which we have data.

M. Fortis and M. Carminati, *The Automatic Packaging Machinery Sector in Italy and Germany*, SpringerBriefs in Business,
DOI 10.1007/978-3-319-12763-7_8

In the Emilia district the 4 biggest companies are Coesia-Seragnoli, IMA, SACMI and Marchesini. In 2010 their total turnover amounted to 1.9 billion euros out of a total of 3.1 billion euros for the entire district.

In the German district there are 7 major companies (Bosch Packaging Technology, Optima, Bausch & Ströbel, Gerhard Schubert, Groninger, Kocher-plastik and R. Weiss). In 2009 (the most complete data we have) their total sales amounted to almost 1.3 billion euros. This does not include sales for R. Weiss, but even taking this into account we can safely say that it is unlikely to reach the figure for the 4 biggest Bologna companies, especially given the fact that of the 7 firms R. Weiss is the smallest employer.

Thus the biggest Italian firms have higher revenues than the biggest German companies.

As the Fig. 8.1 shows, in 2010:

- The biggest Italian company in the packaging machinery industry, Coesia Seragnoli, has revenues of 871 million euros against 717 million of Bosch packaging Technology, which is the biggest German firm.
- The second Italian company, IMA, has turnover twice more than that of the second German firm, Optima: 503 million euros against 226 million.
- The third Italian firm, SACMI, has a turnover of 383 million euros against Gerhard Schubert's 113 million.
- The fourth Italian company, Marchesini, has revenues of 180 million euros compared with 105 million euros for Bausch & Ströbel.

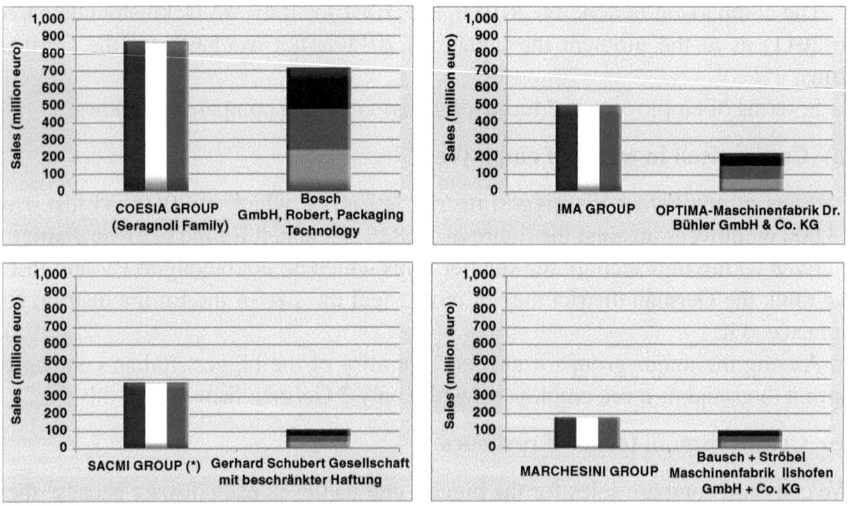

(*) The data used are those related only to the divisions involved in the packaging sector(Beverage & Packaging Division, Plastic Division, Food Machinery and Inspection Division)

Fig. 8.1 The turnover of the four main Italian and German groups of packaging machinery industry: year 2010. *Source* Compiled by Fondazione Edison on AIDA and AMADEUS Database (Bureau van Dijk 2012a, b) and BOSCH data (Bosch 2011)

References

Bosch—Bosch Packaging Technology (2011) Packed as promises—150 years of packaging Technology. Interpack 2011—Press Conference

Bureau van Dijk-Company information and business intelligence (2012a), AIDA Database

Bureau van Dijk-Company information and business intelligence (2012b), AMADEUS Database

MIX
Papier aus verantwortungsvollen Quellen
Paper from responsible sources
FSC® C105338

If you have any concerns about our products,
you can contact us on
ProductSafety@springernature.com

In case Publisher is established outside the EU,
the EU authorized representative is:
Springer Nature Customer Service Center GmbH
Europaplatz 3, 69115 Heidelberg, Germany

Printed by Libri Plureos GmbH
in Hamburg, Germany